T0185929

Fungi for Human Health

Uzma Azeem • Khalid Rehman Hakeem • M. Ali

Fungi for Human Health

Current Knowledge and Future Perspectives

 Springer

Uzma Azeem
Former Assistant Professor in Botany
Sanmati Government College of Science
Education and Research
Jagraon, Ludhiana, Punjab, India

Khalid Rehman Hakeem
Department of Biological Sciences
Faculty of Science
King Abdulaziz University
Jeddah, Saudi Arabia

M. Ali
Department of Pharmacognosy
Jazan University
Jazan, Saudi Arabia

ISBN 978-3-030-58758-1   ISBN 978-3-030-58756-7   (eBook)
https://doi.org/10.1007/978-3-030-58756-7

This Springer imprint is published by the registered company Springer Nature Switzerland AG
The registered company address is: Gewerbestrasse 11, 6330 Cham, Switzerland

This book is dedicated to our parents and siblings

Preface

This book spotlights macrofungi with health-promising properties, which mainly belong to Basidiomycota (Agaricomycotina) and Ascomycota (Pezizomycotina). Macrofungi or higher fungi are ascomycetous and basidiomycetous mushrooms forming conspicuous, epigeous, or hypogeous sporocarps and are large enough to be seen by the naked eye. The consumption of wild macrofungi by man goes back 13,000 years. Macrofungi vary in structure and reproduction and occur in a wide range of habitats in different ecogeographic zones of the world. Ethnomycological surveys across the globe reveal the food value and therapeutic significance of wild species in diet and folk medicine. Scientific research proves the nutritional and pharmacological properties of these macrofungi. The latter are consecrated with a wealth of nutrients such as carbohydrates, proteins, fats, fatty acids, amino acids, minerals, and vitamins contributing to their food value. The sporocarps/cultured mycelia or cultured broths of these fungi are rich in numerous high and low molecular weight bioactive constituents. These bioactive components include polysaccharides, proteins, fatty acids, proteoglycans, terpenoids, and phenolics, accounting for a broad spectrum of pharmacological activities such as antioxidant, antitumor, antidiabetic, antibacterial, antifungal, immunomodulatory, antimalarial, and antiviral. Several health-promoting products of macrofungal origin are available in market in the form of tablets, capsules, syrups, pastes, and powders. In spite of huge benefits of macrofungi, these are often overlooked as far as conservation efforts are concerned. Many macrofungal species are red listed by IUCN and need immediate attention to conserve and ensure sustainable use of this inexpensive natural treasure with huge health benefits. In this book, we endeavored to highlight the future prospects of macrofungi and tried to shed light on the taxonomy, ecology, ethnomycology, nutraceutical composition, bioactive active and pharmacological activities, commercialization, and conservation. Some information on cultivation and toxigenic macrofungi is also provided. Hopefully, the information cited within this book will prove beneficial to the mycophiles, amateur naturalists, the general public, researchers, and industrialists interested in the consumption, research, and marketing of macrofungi.

Jagraon, Ludhiana, Punjab, India   Uzma Azeem
Jeddah, Saudi Arabia   Khalid Rehman Hakeem
Jazan, Saudi Arabia   M. Ali

Contents

About the Authors

Uzma Azeem, Ph.D., is a postgraduate, M.Phil. and doctorate from Punjabi University, Patiala, Punjab, India. She spent more than a decade there from 2006 to 2018. Dr. Azeem completed her Doctorate in Botany with specialization in Mycology and Plant Pathology from Punjabi University, Patiala, Punjab, India, in 2018. She worked at Sri Guru Granth Sahib World University, Fatehgarh Sahib, Punjab, India, as an Assistant Professor in the Department of Botany from August 1, 2018, to January 21, 2019. Dr. Azeem joined Sanmati Government College of Science Education and Research, Jagraon, Ludhiana, Punjab, India, on January 21, 2019, and worked there for a short period. She has been awarded with the Maulana Azad National Fellowship for Minority Students by the University Grants Commission, New Delhi, India, in 2011. So far, she has authored nine papers (eight research papers plus one review paper) in peer-reviewed national and international journals. Dr. Azeem has participated and presented her research work at various conferences/symposia across India winning appreciation and awards. This is her first endeavor in book publication.

Khalid Rehman Hakeem, Ph.D., is Professor at King Abdulaziz University, Jeddah, Saudi Arabia. After completing his doctorate (Botany, specialization in Plant Eco-physiology and Molecular Biology) from Jamia Hamdard, New Delhi, India, in 2011, he worked as a Lecturer at the University of Kashmir, Srinagar, for a short period. Later, he joined Universiti Putra Malaysia, Selangor, Malaysia, and worked there as Postdoctoral Fellow in 2012 and Fellow Researcher (Associate Prof.) from 2013 to 2016. Dr. Hakeem has more than 10 years of teaching and research experience in plant eco-physiology, biotechnology and molecular biology, medicinal plant research, plant-microbe-soil interactions, as well as environmental studies. He is the recipient of several fellowships at both national and international levels; he has also served as Visiting Scientist at Jinan University, Guangzhou, China. Currently, he is involved in a number of international research projects with different government organizations.

So far, Dr. Hakeem has authored and edited more than 50 books with international publishers, including Springer Nature, Academic Press (Elsevier), and CRC

Press. He also has to his credit more than 110 research publications in peer-reviewed international journals and 60 book chapters in edited volumes with international publishers.

At present, Dr. Hakeem serves as an editorial board member and reviewer of several high-impact international scientific journals from Elsevier, Springer Nature, Taylor and Francis, Cambridge, and John Wiley Publishers. He is included in the advisory board of Cambridge Scholars Publishing, UK. Dr. Hakeem is also a fellow of Plantae group of the American Society of Plant Biologists; member of the World Academy of Sciences; member of the International Society for Development and Sustainability, Japan; and member of Asian Federation of Biotechnology, Korea. Dr. Hakeem has been listed in Marquis Who's Who in the World between 2014 and 2020. Currently, Dr. Hakeem is engaged in studying plant processes at eco-physiological as well as molecular levels.

M. Ali is currently working under the Ministry of Higher Education as a Lecturer of Pharmacognosy in the Department of Pharmacognosy, College of Pharmacy, Jazan University, Kingdom of Saudi Arabia (KSA). He received his Ph.D. from Jamia Hamdard University, New Delhi, India, in 2012 for "Phytochemical and Pharmacological Studies of an Anticancer Medicinal Plant and Its Authentication Using Molecular Biology Techniques." Dr. Ali is also recipient of Government of India's Junior Research Fellowship (JRF, GATE) for pursuing a master's degree in pharmacy (M. Pharm) for "Enhanced Production of Vasicine and Vasicinone from Callus Culture of Adhatoda vasica." He has published/presented about 25 research articles in national and international journals of repute. Dr. Ali has reviewed scientific papers in the field of pharmaceutical science.

Chapter 1
Introduction

Fungi constitute a group of eukaryotic, spore-bearing, achlorophyllous, saprotrophic or parasitic, sexually or asexually reproducing organisms with unicellular to filamentous (hyphal) forms, filaments, or hyphae branched or unbranched typically enclosed by cell wall made up of chitin or cellulose or both (Alexopoulos et al. 1996). In the sixteenth century, fungi were represented by two genera, *Fungus* and *Tuber* (De Lobel 1581). At the end of the seventeenth century and start of the eighteenth century, De Tournefort (1656–1708), father of modern generic concept, added five new genera including *Agaricus* (Bauhin 1623; De Tournefort 1694, 1700). Great mycologist Micheli, known as the father of mycology, proposed 30 new genera to fungi (Micheli 1729). This number is incessantly piling up with new reports. Approximately, 1200 new species are being reported per annum. Fungi are classified into six phyla: *Ascomycota*, *Basidiomycota*, *Chytridiomycota*, *Glomeromycota*, *Microsporidia*, and *Zygomycota* (Kirk et al. 2008). Based on morphology, ecology, phylogeny, and various other extrapolation techniques, fungal diversity is estimated to be approximately 3 to 5.1 million globally (Dai 2010; Blackwell 2011; Hawksworth 2012). However, 92% of these fungi are still undescribed (Hawksworth and Lücking 2017). Majority of all the described fungi belong to the subkingdom *Dikarya* including *Ascomycota* and *Basidiomycota*. *Ascomycota* represents the largest phylum of kingdom fungi and consists of nearly 90,000 described species of fungi belonging to *Taphrinomycotina* (yeast-like and some filamentous fungi), *Saccharomycotina* (the true yeasts), and *Pezizomycotina* (with majority of the filamentous and mushroom-forming ascomycetes). *Basidiomycota* follows *Ascomycota* and comprises of approximately 50,000 species of fungi divided into *Pucciniomycotina* (rust fungi: plant pathogens), *Ustilaginomycotina* (true smuts, some yeasts, and some filamentous fungi), and *Agaricomycotina* (most of the mushroom-producing fungi) (Cannon et al. 2018; Niskanen et al. 2018). The mushroom-forming fungi, macrofungi, constitute a phylogenetically heterogeneous fungal group with species forming epigeal/hypogeal, large spore-bearing structures called sporocarps. An estimated number of macrofungi in the world have risen from

140,000 to 1,250,000 with majority of these belonging to *Basidiomycota* (*Agaricomycotina*) followed by *Ascomycota* (*Pezizomycotina*) and *Zygomycota* (a few taxa occasionally) including species with health-promising effects. Macrofungi include true mushrooms, poricins, earthstars, club corals, pig's ears, false truffles, stinkhorns, coral fungi, crust fungi, polypores, conks, brittle gills, milk caps, earth fans, jelly fungi, true truffles, morels, earth tongues, chicken lips, green elf cup fungi, and toadstools (Hawksworth 2001; Mueller et al. 2007; Tripathi et al. 2017; Thiers and Halling 2018). This number keeps on increasing with the addition of new taxa by reports coming from less explored and unexplored regions across the globe. The use of edible macrofungi as human food is 13,000 years back as indicated by archaeological records from Chile (Rojas and Mansur 1995). Extreme levels of mycophilia have been observed in most regions of Southern and Eastern Europe, Turkey, parts of Africa, Mexico, and majority of the Asian continent (Boa 2004). Macrofungi show structural and ecological variations and occur in almost all eco-geographic zones throughout the world. These are heterotrophic organisms with saprotrophic, symbiotic, or parasitic mode of nutrition (Naranjo-Ortiz and Gabaldón 2019). In spite of great benefits of macrofungi to human health, these are often neglected and misunderstood organisms (De Mattos-Shipley et al. 2016; Tripathi et al. 2017; Roncero-Ramos and Delgado-Andrade 2017). However, ethnomycological surveys and scientific research across the world successfully changed these misconceptions and unveil the significance of macrofungi as food and medicine. Edible macrofungi are eaten as food, medicinal species are used for the treatment of various human disorders, while some macrofungal species are both edible and medicinal. Some have value in religious rituals, crafts, and myths and beliefs (Dugan 2011). Macrofungi are rich in diverse nutrients such as carbohydrates, proteins, fats, dietary fibers, amino acids, minerals, and vitamins but low in fats and calories contributing towards their food value (Turfan et al. 2018). More recently several species are used as dietary supplements (DSs), prebiotics, and functional foods (Vikineswary and Chang 2013; Nowak et al. 2018; Perera et al. 2018; Üstün et al. 2018). The essences and extracts prepared from medicinal species of macrofungi have been consumed as alternative medicine in China, Japan, Korea, and Eastern Europe for hundreds of years (Lakhanpal and Rana 2005). Scientific research proves that the extracts made from sporocarps/cultured mycelia or cultured broths of medicinal macrofungi exhibit a number of pharmacological activities such as antioxidant, antidiabetic, antitumor, antimicrobial, anti-inflammatory, immunomodulatory, neuroprotective, antimalarial, antiviral, etc. which can be associated with their unique mycochemical composition. Macrofungi are reservoirs of several bioactive constituents including polysaccharides, proteins, polyphenols, terpenoids, steroids, polyketides, etc. This supports their candidature as natural resources for novel drug discovery (Wasser 2017; Chang and Wasser 2018; Elkhateeb 2020). During the past few decades, much attention is paid on exploring the natural resources for drug development against various diseases in an attempt to get rid of many precarious health effects of synthetic drugs (Karimi et al. 2015; Sorokina and Steinbeck 2020). New findings in the field of medical mycology enhance knowledge base and dimensions of this subject. Keeping in view the growing emphasis in exploring the natural resources with health-promising effects as nutraceuticals and pharmaceuticals, we endeavored to highlight the significance of macrofungi in human health.

References

Alexopoulos CJ, Mims CW, Blackwell M (1996) Introductory mycology, 4th edn. Wiley, New York

Bauhin C (1623) Pinax theatri botanici. Sumptibus et typis Ludovici Regis, Basileae Helvet

Blackwell M (2011) The Fungi: 1, 2, 3 … 5.1 million species? Am J Bot 98(3):426–438

Boa E (2004) Wild edible fungi: a global overview of their use and importance to people. Non-wood forest products (No. 17). FAO, Rome

Cannon PF, Aguirre-Hudson B, Aime C, Aime MC, Vanzela A, Ainsworth AM, Bidartondo M, Gaya E, Hawksworth D, Kirk PM, Leitch I, Lucking R (2018) Definition and diversity. In: Willis KJ (ed) State of the world's fungi. Royal Botanic Gardens, Kew, pp 4–11

Chang ST, Wasser SP (2018) Current and future research trends in agricultural and biomedical applications of medicinal mushrooms and mushroom products (review). Int J Med Mushrooms 20(12):1121–1133

Dai YC (2010) Hymenochaetaceae (Basidiomycota) in China. Fungal Divers 45(1):131–343

De Lobel M (1581) Kruydtboeck oft Beschryvinghe van allerleye ghewassen, kruyderen, hesteren, ende gheboomten. Antwerpen, By Chrostoffel Plantyn

De Mattos-Shipley KMJ, Ford KL, Alberti F, Banks AM, Bailey AM, Foster GD (2016) The good, the bad and the tasty: the many roles of mushrooms. Stud Mycol 85:125–157

De Tournefort JP (1694) Élémens de botanique: ou méthode pour connoitre les plantes, vol 1. De L'Imprimerie Royale, Paris

De Tournefort JP (1700) Institutiones rei herbariae, vol 1. E typographia regia, Paris

Dugan FM (2011) Conspectus of world ethnomycology: fungi in ceremonies, crafts, diets, medicines and myths. St. Paul, American Phytopathological Society

Elkhateeb WA (2020) What medicinal mushroom can do? Chem Res J 5(1):106–118

Hawksworth DL (2001) The magnitude of fungal diversity: the 1.5 million species estimate revisited. Microbiol Res 105(12):1422–1432

Hawksworth DL (2012) Global species number of fungi: are tropical studies and molecular approaches contributing to a more robust estimate? Biodivers Conserv 21(9):2425–2433

Hawksworth DL, Lücking R (2017) Fungal diversity revisited: 2.2 to 3.8 million species. In: Heitman J, Howlett BJ, Crous PW, Stukenbrock EH, James TY, Gow NAR (eds) The fungal kingdom. American Society for Microbiology, Washington, DC, pp 79–95

Karimi A, Majlesi M, Rafieian-Kopaei M (2015) Herbal versus synthetic drugs: beliefs and facts. J Nephropharmacol 4(1):27–30

Kirk PM, Cannon PF, Minter DW, Stalpers JA (2008) Ainsworth and Bisby's dictionary of the fungi, 10th edn. CABI, Wallingford

Lakhanpal TN, Rana M (2005) Medicinal and nutraceutical genetic resources of mushrooms. Plant Genet Resours 3(2):288–303

Micheli PA (1729) Nova plantarum genera juxta Tournefortianam methodum disposita. Typis Bernardi Paperinii, Florence

Mueller GM, Schmit JP, Leacock PR, Buyck B, Cifuentes J, Desjardin DE, Halling RE, Hjortstam K, Iturriaga T, Larsson KH, Lodge DJ (2007) Global diversity and distribution of macrofungi. Biodivers Conserv 16(1):37–48

Naranjo-Ortiz MA, Gabaldón T (2019) Fungal evolution: diversity, taxonomy and phylogeny of the fungi. Biol Rev 94(6):2101–2137

Niskanen T, Douglas B, Kirk P, Crous PW, Lücking R, Matheny PB, Cai L, Hyde K (2018) New discoveries: species of fungi described in 2017. In: Willis KJ (ed) State of the world's fungi. Royal Botanic Gardens, Kew, pp 18–23

Nowak R, Nowacka-Jechalke N, Juda M, Malm A (2018) The preliminary study of prebiotic potential of Polish wild mushroom polysaccharides: the stimulation effect on *Lactobacillus* strains growth. Eur J Nutr 57(4):1511–1521

Perera N, Yang F-L, Lu Y-T, Li L-H, Hua K-F, Wu S-H (2018) *Antrodia cinnamomea* galactomannan elicits immuno-stimulatory activity through toll-like receptor. Int J Biol Sci 14(10):1378–1388

Rojas C, Mansur E (1995) Ecuador: informaciones generales sobre productos non madereros en Ecuador. In: Memoria, consulta de expertos sobre productos forestales no madereros para America Latina y el Caribe, Serie Forestal. FAO Regional Office for Latin America and the Caribbean, Santiago, pp 208–223

Roncero-Ramos I, Delgado-Andrade C (2017) The beneficial role of edible mushrooms in human health. Curr Opin Food Sci 14:122–128

Sorokina M, Steinbeck C (2020) Review on natural products databases: where to find data in 2020. J Cheminform 12(20):1–51

Thiers BM, Halling RE (2018) The macrofungi collection consortium. Appl Plant Sci 6(2):1–7

Tripathi NN, Singh P, Vishwakarma P (2017) Biodiversity of macrofungi with special reference to edible forms: a review. J Indian Bot Soc 96(3):144–187

Turfan N, Pekşen A, Kibar B, Ünal S (2018) Determination of nutritional and bioactive properties in some selected wild growing and cultivated mushrooms from Turkey. Acta Sci Pol Hortorum Cultus 17(3):57–72

Üstün NŞ, Bulam S, Pekşen A (2018) The use of mushrooms and their extracts and compounds in functional foods and nutraceuticals. Türkmen, A (ed.) 1:1205–1222

Vikineswary S, Chang ST (2013) Edible and medicinal mushrooms for sub-health intervention and prevention of lifestyle diseases. Tech Monitor 3:33–43

Wasser SP (2017) Medicinal mushrooms in human clinical studies. part 1. anticancer, oncoimmu-nological, and immunomodulatory activities: a review. Int J Med Mushrooms 19(4):279–317

Chapter 2
Taxonomy

Fungi constitute the third kingdom of organisms which are highly diverse, heterotrophic eukaryotes characterized by chitinous cell wall ranging from unicellular to syncytial filamentous forms divided into six phyla *Ascomycota*, *Basidiomycota*, *Chytridiomycota*, *Glomeromycota*, *Microsporidia*, and *Zygomycota* (Kirk et al. 2008). Majority of all the described fungi belong to the subkingdom *Dikarya* represented by *Ascomycota*, *Taphrinomycotina* (yeast-like and some filamentous fungi), *Saccharomycotina* (the true yeasts), and *Pezizomycotina* (with majority of the filamentous and mushroom-forming ascomycetes), and *Basidiomycota*, *Pucciniomycotina* (rust fungi-plant pathogens), *Ustilaginomycotina* (true smuts, some yeasts, and some filamentous fungi), and *Agaricomycotina* (most of the mushroom-producing fungi) (Cannon et al. 2018; Niskanen et al. 2018). Macrofungi also known as higher fungi/macromycetes include the taxa known to form large, conspicuous, and epigeous or hypogeous sporocarps which are the "fruit" or the reproductive parts of the vegetative fungal mycelium. Sporocarps are the products of sexual reproduction (Fig. 2.1) (Aneja and Mehrotra 2015). There are approximately 1,250,000 macrofungi with majority belonging to *Basidiomycota* (*Agaricomycotina*: *Tremellomycetes*, *Dacrymycetes*, and *Agaricomycetes*) followed by *Ascomycota* (*Pezizomycotina*: *Leotiomycetes*, *Pezizomycetes*, and *Sordariomycetes*) and *Zygomycota* (a few in number). The class *Agaricomycetes* dominates with maximum number of macrofungi among all the ascomycetous and basidiomycetous classes consisting of macrofungi (Mueller et al. 2007; Tripathi et al. 2017; Thiers and Halling 2018). In *Agaricomycotina*, the yeast forming states occur in *Tremellomycetes*. The latter includes yeast forms that can be dimorphic, forming large gelatinous basidiocarps (e.g., *Cryptococcus*, *Tremella*, *Cystofilobasidium*) or species with no yeast stage (Hibbett 2006; Hibbett et al. 2007; Adl et al. 2012, 2018). *Dacrymycetes* (jelly fungi) include the wood-degrading, gelatinous species with large highly pigmented basidiocarps (Hibbett 2006; Shirouzu et al. 2013, 2016). The class *Agaricomycetes* represents the largest and the most diverse group in *Agaricomycotina*. It includes nearly 21,000 described species

U. Azeem et al., *Fungi for Human Health*, https://doi.org/10.1007/978-3-030-58756-7_2

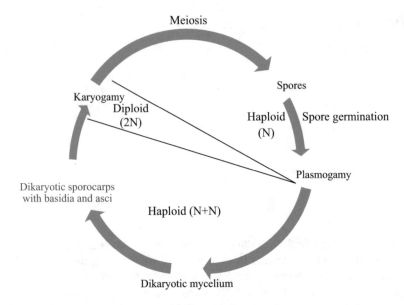

Fig. 2.1 Production of sporocarp in sexual life cycle of a macrofungus

(Kirk et al. 2008). The members of the class produce a variety of forms of basidio-carps. The class includes 16,000 identified species of mushroom-forming fungi (Petersen 2012; Money 2016) classified in *Agaricales* (true mushrooms), *Atheliales*, *Boletales* (poricins), *Geastrales* (earth stars), *Gomphales* (club corals, pigs ears), *Hysterangiales* (false truffles), *Phallales* (stinkhorns), *Cantharellales* (coral fungi), *Corticiales* (crust fungi), *Hymenochaetales* (crust fungi), *Polyporales* (polypores, conks), *Russulales* (brittle gills, milk caps), *Thelephorales* (earth fans), and *Trechisporales* (Thiers and Halling 2018). The structure of the basidiocarps is very complex with spore-bearing surface either on lamellae or inside the tubes opening through pores. The lamellate species are known as agarics/euagarics (*Agaricales*) and the poroid species with tubes as boletes and polypores. A typical agaric basidi-ocarp consists of five parts: the pileus, lamellae (gills), annulus (ring), stipe, and volva (Fig. 2.2). Various forms of basidiocarps and hymenophore configurations are found among different orders of the class *Agaricomycetes*, e.g., resupinate, effused-reflexed, pileate-stipitate/substipitate or sessile, coralloid, polyporoid, corticioid, and gasteroid (Hibbett 2007). The resupinate forms occur in almost all orders of *Agaricomycetes*. The orders, namely, *Geastrales*, *Hysterangiales*, and *Phallales*, do not consist of any resupinate form. The orders *Amylocorticiales*, *Atheliales*, *Corticiales*, *Jaapiales*, *Trechisporales*, and *Lepidostromatales* comprise only or almost only resupinate forms. The rest of the members belonging to different orders of the class *Agaricomycetes* include resupinate to pileate-stipitate/sessile forms (epigeous or hypogeous, gilled or not, and gasteroid or not). Several genera classi-fied in different of orders of the class include macrofungal species known to form large conspicuous basidiocarps of various kinds, e.g., *Agaricus*, *Amanita*, *Clavaria*,

Pileus/Cap

Gills/Lamellae

Annulus/Ring

Stipe/Stalk

Volva

Fig. 2.2 Morphology of a typical agaric mushroom (e.g. *Amanita phylloides*)

Coprinopsis, Coprinus, Entoloma, Fistulina, Lepiota, Lycoperdon, and *Pleurotus* (*Agaricales*); *Hydnum* and *Cantharellus* (*Cantharellales*); *Auricularia auricula-judae* (wood ear) (*Auriculariales*); *Boletus* and *Paxillus* (*Boletales*); *Geastrum* (*Geastrales*); *Gomphus* and *Ramaria* (*Gomphales*); *Phallus* (*Phallales*); *Phellinus* and *Trichaptum* (*Hymenochaetales*); *Fomes, Ganoderma, Lenzites,* and *Sparassis* (Polyporales); Thelephora (*Thelephorales*); and *Russula, Lactarius,* and *Hericium* (Russulales). The hymenophore of these species exhibits huge variation in configuration from smooth to hydnoid, poroid, toothed, spiny, tuberculate, etc. (Hibbett et al. 2014). Several types of fruiting bodies called as ascocarps are formed in *Ascomycota* such as apothecia, perithecia, pseudoperithecia, and cleistothecia (Pöggeler et al. 2006; Schmitt 2011). Macrofungi such as morels, false morels, earth tongues, truffles, chicken lips, green elf cup species belong to *Ascomycota* (*Pezizomycotina*) spreading over *Leotiales*, e.g., *Leotia, Chlorociboria,* Pezizales, e.g., *Morchella, Geoglossum, Gyromitra, Helvella, Peziza, Pyronema, Tuber,* and *Verpa* and *Hypocreales,* e.g., *Cordyceps* spp. with diverse forms of ascocarps (Adl et al. 2012; Thiers and Halling 2018). In field, one can identify the sporocarps (basidiocarps and ascocarps) with their diverse color, shape, dimensions, and consistency. These appear fleshy, subfleshy/leathery or sometimes woody hard growing singly or in clusters on a vast diversity of substrates as saprotrophs, parasites, or symbionts (Fig. 2.3). Microscopically, the sporocarp is made up of mycelium consisting of hyphae. The ascospore-bearing structures are asci within the ascocarps and basidiospores are produced exogenously attached to the sterigmata of the club-shaped basidia. The ascospores and basidiospores are usually eight per ascus and four per basidium, respectively. The number of spores varies among different taxa, but this is constant for a species. The spores vary in color, size, and shape and are thin to thick walled. The spore wall may be smooth or ornamented variously and an

Fig. 2.3 The sporocarps of wild (edible/medicinal) macrofungi growing (singly or in clusters) as saprotrophs (in soil, grass, animal dung, leaf litter, woody debris, decaying trees), tree parasites or symbionts (ECM forms and termite-symbionts) distributed in different parts of the world (1) *Agaricus campestris* (2) *Amanita caesarea* (3) *A. vaginata* (4) *Auricularia auricula-judae*

Fig. 2.3 (continued) (5) *Armillaria mellea* (6) *Boletus pinophilus* (7) *Calvatia gigantea* (8) *Cookeina speciosa* (9) *Coprinus comatus* (young and old fruit bodies) (10) *Cordyceps militaris* (11) *Fomes fomentarius* (12) *Ganoderma lucidum* (13) *Grifola frondosa* (14) *Gyromitra esculenta* (15) *Helvella crispa* (16) *Hericium erinaceus* (17) *Hydnum repandum* (18) *Hygrocybe miniata* (19) *Hygrophorus chrysodon* (20) *Inonotus obliquus* (21) *Laccaria laccata* (22) *Lactarius deliciosus* (23) *Lactifluus pelliculatus* (24) *Leccinum aurantiacum* (25) *Lentinula edodes* (26) *Leucopaxillus giganteus* (27) *Lycoperdon perlatum* (28) *Lycoperdon pratense* (29) *Marasmius buzungolo* (30) *Morchella esculenta* (31) *Phellinus linteus* (32) *Pleurotus cornucopiae* (33) *P. ostreatus* (34) *Psathyrella tuberculata* (35) *Ramaria botrytis* (36) *Russula sesenagula* (37) *Sarcodon imbricatus* (38) *Schizophyllum commune* (39) *Sparassis crispa* (40) *Terfezia arenaria* (41) *Termitomyces microcarpus* (42) *T. titanicus* (43) *Trametes versicolor* (44) *Tremella mesenterica* (45) *Volvariella volvacea* (47) *Xerocomus cisalpinus* (47) *Xylaria hypoxylon*. (https://en.wikipedia.org; http://www.asociacionmicologicamairei.com; https://www.efta-online.org; https://www.first-nature.com; https://www.flickr.com; https://www.mushroomexpert.com; https://www.pinterest.com; www.medicalmushrooms.net; www.muse.it; www.tropicalfungi.org)

important taxonomic feature. Several sterile structures are also found in the hymenium, subhymenium, or tramal region (Alexopoulos et al. 1996; Hibbett et al. 2014). Literature reports reveal that several macrofungal taxa have immense significance in human health as food and medicine. These health benefitting macrofungi are classified mainly in *Agaricomycotina* (*Tremellomycetes, Dacrymycetes, Agaricomycetes*) followed by *Pezizomycotina* (*Leotiomycetes, Pezizomycetes, Sordariomycetes*). Majority of the edible and medicinal macrofungi occur in *Agaricomycetes*, e.g., *Agaricus, Boletus, Fomes, Ganoderma, Lycoperdon, Pleurotus, Russula*, etc. (Tripathi et al. 2017). However, some agaricomycetous species are toadstools, e.g., *Amanita phalloides, Galerina autumnalis*, and others, which sometimes might cause severe toxicity (Benjamin 1995).

References

Adl SM, Simpson AG, Lane CE, Luke's J, Bass D, Bowser SS, Brown M, Burki F, Dunthorn M, Hampl V, Heiss A, Hoppenrath M, Lara E, Le Gall L, Lynn DH, McManus H, Mitchell EAD, Mozley-Stanridge SE, Parfrey LW, Pawlowski J, Rueckert S, Shadwick L, Schoch CL, Smirnov A, Spiegel FW (2012) The revised classification of eukaryotes. J Eukaryot Microbiol 59(5):1–45

Adl SM, Bass D, Lane CE, Lukes J, Schoch CL, Smirnov A, Agatha S, Berney C, Brown MW, Burki F, Cardenas P, Cepicka I, Chistyakova L, del Campo J, Dunthorn M et al (2018) Revisions to the classification, nomenclature and diversity of eukaryotes. J Eukaryot Microbiol 66(1):4–119

Alexopoulos CJ, Mims CW, Blackwell M (1996) Introductory mycology, 4th edn. Wiley, New York

Aneja KR, Mehrotra RS (2015) An introduction to mycology, 2nd edn. New Age International Pvt. Ltd., Daryaganj, New Delhi

Benjamin DR (1995) Mushrooms: poisons and panaceas. WH Freeman, New York

Cannon PF, Aguirre-Hudson B, Aime C, Aime MC, Vanzela A, Ainsworth AM, Bidartondo M, Gaya E, Hawksworth D, Kirk PM, Leitch I, Lucking R (2018) Definition and diversity. In: Willis KJ (ed) State of the world's fungi. Royal Botanic Gardens, Kew, pp 4–11

Hibbett DS (2006) A phylogenetic overview of the Agaricomycotina. Mycologia 98(6):917–925

Hibbett DS (2007) After the gold rush, or before the flood? Evolutionary morphology of mushroom forming fungi (Agaricomycetes) in the early 21st century. Mycol Res 111:1001–1018

Hibbett DS, Binder M, Bischoff JF, Blackwell M, Cannon PF, Eriksson OE, Huhndorf S, James T, Kirk PM, Lucking R, Lumbsch HT, Lutzoni F, Matheny PB, Mclaughlin DJ, Powell MJ, Redhead S, Schoch CL, Spatafora JW, Zhang N (2007) A higher-level phylogenetic classification of the fungi. Mycol Res 111(5):509–547

Hibbett DS, Bauer R, Binder M, Giachini AJ, Hosaka K, Justo A, Larsson E, Larsson KH, Lawrey JD, Miettinen O, Nagy LG (2014) Agaricomycetes. In: Systematics and evolution. Springer, Berlin/Heidelberg, pp 373–429

Kirk PM, Cannon PF, Minter DW, Stalpers JA (2008) Ainsworth and Bisby's dictionary of the fungi, 10th edn. CABI, Wallingford

Money NP (2016) Fungal diversity. In: The fungi. Academic Press, pp 1–36

Mueller GM, Schmit JP, Leacock PR, Buyck B, Cifuentes J, Desjardin DE, Halling RE, Hjortstam K, Iturriaga T, Larsson KH, Lodge DJ (2007) Global diversity and distribution of macrofungi. Biodivers Conserv 16(1):37–48

Niskanen T, Douglas B, Kirk P, Crous PW, Lücking R, Matheny PB, Cai L, Hyde K (2018) New discoveries: species of fungi described in 2017. In: Willis KJ (ed) State of the world's fungi. Royal Botanic Gardens, Kew, pp 18–23

Petersen JH (2012) The kingdom of fungi. Princeton University Press, Princeton, NJ

Pöggeler S, Nowrousian M, Kück U (2006) Fruiting-body development in ascomycetes. In: Growth, differentiation and sexuality. Springer, Berlin/Heidelberg, pp 325–355

Schmitt I (2011) Fruiting body evolution in the ascomycota: a molecular perspective integrating lichenized and non-lichenized groups. In: Evolution of fungi and fungal-like organisms. Springer, Berlin/Heidelberg, pp 187–204

Shirouzu T, Hirose D, Oberwinkler F, Shimomura N, Maekawa N, Tokumasu S (2013) Combined molecular and morphological data for improving phylogenetic hypothesis in Dacrymycetes. Mycologia 105(5):1110–1125

Shirouzu T, Uno K, Hosaka K, Hosoya T (2016) Early-diverging wood-decaying fungi detected using three complementary sampling methods. Mol Phylogenet Evol 98:11–20

Thiers BM, Halling RE (2018) The macrofungi collection consortium. Appli Plant Sci 6(2):1–7

Tripathi NN, Singh P, Vishwakarma P (2017) Biodiversity of macrofungi with special reference to edible forms: a review. J Indian Bot Soc 96(3 and 4):144–187

Websites Followed

http://www.asociacionmicologicamairei.com
https://en.wikipedia.org
https://www.efta-online.org
https://www.first-nature.com
https://www.flickr.com
https://www.mushroomexpert.com
https://www.pinterest.com
www.medicalmushrooms.net
www.muse.it
www.tropicalfungi.org

Chapter 3
Ecology and Distribution

Macrofungi are ubiquitous in distribution and occur in various regions across the world such as in North America, tropical (Central and South America), temperate South America, Africa, temperate and tropical Asia, Australasia (Australia, New Caledonia, New Guinea, and New Zealand), and Western Europe (Mueller et al. 2007). As per the estimate of the Macrofungi Collection Consortium (MaCC) project (2012–2017) financially supported by the National Science Foundation (NSA), United States (US), there are 1,250,000 specimens of macrofungi from North America (73%); Central America, South America, and the Caribbean region (18%); Europe (5%); Asia and the Pacific region (2%); Australia and New Zealand (1%); and Africa (1%). *Polyporaceae* is the dominant family with 202,069 specimens followed by *Agaricaceae* (100,416), *Russulaceae* (85,127), *Tricholomataceae* (77,078), *Strophariaceae* (65,111), *Boletaceae* (59,781), *Cortinariaceae* (55,833), *Hymenochaetaceae* (40,564), *Stereaceae* (37,607), *Amanitaceae* (37,105), *Mycenaceae* (36,376), *Meruliaceae* (35,599), *Inocybaceae* (33,042), *Hygrophoraceae* (28,253), *Fomitopsidaceae* (26,479), *Corticiaceae* (25,137), *Entolomataceae* (24,826), *Gomphaceae* (20,373), *Psathyrellaceae* (18,558), and *Peniophoraceae* (17,585) (Thiers and Halling 2018). The number of known and unknown macrofungal species is continuously increasing as there are still underexplored and unexplored regions in the world. It has increased from 140,000 (Hawksworth 2001) to 1,250,000 (Thiers and Halling 2018) and is still continuous to increase. Macrofungi have a wide distribution range and occur in almost all forest types across the globe such as moist, dry, deciduous, evergreen forests of tropical, subtropical, temperate, and alpine regions (Tripathi et al. 2017; Sarrionandia and Salcedo 2018; Monteiro et al. 2020). Fungal diversity flourishes more and is higher in tropics than the temperate regions. The reason behind this is the suitable environmental conditions prevailing across the year, diverse forms of vascular plants providing habitat to macrofungi and the occurrence of various ecotones in tropical forests (Hawksworth 2001; Kark 2007). As accessed by Lopez-Quintero et al. (2012), tropical lowland forests in Colombia are a home for 403 macromycetes

© The Editor(s) (if applicable) and The Author(s), under exclusive license
to Springer Nature Switzerland AG 2020
U. Azeem et al., *Fungi for Human Health*,
https://doi.org/10.1007/978-3-030-58756-7_3

spreading over 129 genera and 48 families of *Basidiomycota* and *Ascomycota* with dominancy of *Polyporaceae*, *Marasmiaceae*, and *Agaricaceae*. The gasteroid fungi, 47 species characterized by closed hymenium with 48% *Geastrum* species, have been reported from 3 forest ecosystems (lowland rain forest, *Araucaria* forest, and semi-deciduous forest) in South Brazilian Atlantic forest. Other species found growing in these ecosystems belong to the genera *Bovista*, *Clavatia*, *Cyathus*, *Lycoperdon*, *Morganella*, *Mutinus*, *Phallus*, *Scleroderma*, and *Vascellum* (Alves et al. 2018). Macrofungi belonging to *Agaricomycotina* and *Pezizomycotina* spreading across 8 orders, 24 families, 47 genera, and 57 species have been collected and identified from Tian Shan mountain forest ranges in Kyrgyzstan (Central Asia). *Agaricomycotina* exhibits maximum species diversity with dominant species in *Polyporaceae* (eight species), *Agaricaceae* (six species), *Hymenogastraceae*, and *Psathyrellaceae* (four species) (Cho et al. 2019). Fernández et al. (2020) reported 189 epigeous macrofungal species from the Mediterranean forest ecosystems with majority of the *Basidiomycetes* (*Tricholomataceae*, *Cortinariaceae*, and *Russulaceae*). Approximately, 280 macrofungal species have been reported from the hilly regions of Jammu and Kashmir (India). Majority of these species belong to *Basidiomycota* with the dominancy of macromycetes in the family *Russulaceae* (29 species) followed by *Cortinariaceae* (17 species) and *Boletaceae* (17 species), *Coprinaceae* (12 species), and *Tricholomataceae* (11 species). *Russula* dominate the genera with 21 species followed by *Amanita* (11 species). Among *Ascomycetes*, the fungal diversity follows the trend of *Morchellaceae* (seven species) > *Helvellaceae* (six species) > *Pyronemataceae* (five species) with five species each of *Morchella* and *Helvella* (Wani et al. 2020). The mangrove forest ecosystems distributed along the coastlines in tropics and subtropics also provide shelter to macrofungi. Ghate and Sridhar (2016) reported 46 macrofungal species belonging to 27 genera from 5 mangroves of Southwest India (Paduhithlu, Nadikudru, Sasihithlu, Thokottu, and Batapady). These macrofungi include maximum number of woody litter decomposing species (26) followed by 22 terricolous species and 7 species growing on leaf litter. Tripathi et al. (2017) provided a comprehensive data (1921–2016) of macrofungi including a treasure of health promising edible/medicinal species distributed in a broad spectrum of forest types such as moist, dry, deciduous, evergreen, tropical, subtropical, temperate, and alpine forests across the globe, growing on a variety of substrates/hosts along with the mycologists who reported these taxa. Priyamvada et al. (2017) recorded 113 macrofungal species from the tropical dry evergreen forests of Southern India belonging to 54 genera and 23 families with the dominancy of *Ascomycota* (96%) followed by *Basidiomycota* (4%). The *Agaricaceae* (25.3%) members dominated the mycobiota followed by *Polyporaceae* (15.3%) and *Marasmiaceae* (10.8%). Fungi in large number approximately 193 taxa have been recorded from the Italian Alps (Jamoni 2008). In spite of harsh weather conditions (heavy rains, snow, freezing and thawing, drought and strong sun rays), alpine regions of Switzerland, France, and Germany provide home to diverse macrofungi. Favre (1955) reported 204 species of macrofungi from Switzerland. Majority of these species belong to *Basidiomycota* and only 12 are ascomycetous. These macromycetes include ectomycorrhizal (ECM) taxa symbiotic with alpine plant hosts and saprotrophic species deconstructing the plant litter and organic matter present in soil. Combined list of alpine fungi from Switzerland, France, and Germany including Favre (1955) and the "Swissfungi" database increase the number to 210 ECM taxa with dominancy of *Inocybe* followed by *Cortinarius* (Brunner et al. 2017). Macrofungi are also capable of inhabiting arid and semi-arid regions surrounding the Mediterranean and the Middle East (Iddison 2011). The "desert truffles" (*Pezizales*) are the representatives of such fungi, e.g., *Terfezia, Tirmania, Tuber, Delastria, Loculotuber,* and *Picoa* (Owaid 2018). Macrofungal species, viz., *Helvella acetabulum, H. corium, H. queletii,* and *H. macropus,* have also been found fruiting in the cold desert of Leh, Ladakh, Jammu, and Kashmir, India (Dorjey et al. 2013). Most of the taxonomic reports from different countries have been published on macrofungal diversity in forests, national parks neglecting the urban areas (Borkar et al. 2015; Jang et al. 2016; Kim et al. 2015; Lee et al. 2014; Semwal et al. 2018). Cho et al. (2020) collected and identified macrofungi classified in 37 families, 90 genera, and 139 species form the urbanized areas in Gyeongsang Province located in the southern part of Korea. Of the total macrofungal diversity, 37 species have been found growing on legally protected trees, 48 species in urban parks, and 84 species in cultural heritage protection zones. Macrofungi exhibit seasonal variation in growth and development with maximum growth in the rainy season because of high moisture and nutrient availability needed for spore germination, growth, and development of the sporocarps (Andrew et al. 2013; Kim et al. 2017; Thulasinathan et al. 2018). Macrofungi are achlorophyllous, do not perform photosynthesis, and are heterotrophs. These play diverse roles in forest ecosystems such as nutrient cycling, as saprotrophs, predators, pathogens, and parasites, and affect the diversity of other forest communities (Bajpai et al. 2019; Copoț and Tănase 2019). The macrofungal species can be saprotrophs (humicolous, lignicolous, wood deconstructors, coprophilous, and saprotrophs on litter), symbiotic, or parasitic growing solitary/in groups at one place or scattered (singly or in clusters) on a variety of substrates including different types of soil rich in moisture and humus, grassy ground, dead decaying wood and organic matter, leaf litter, animal dung, and termite mounds, on a broad spectrum of live trees (Naranjo-Ortiz and Gabaldón 2019) and on plastic debris such as *Agaricus bisporus* and *Marasmius oreades* (Brunner et al. 2018). Maximum species generally utilize tree trunks as substrates followed by soil, leaves, twigs, and the minimum number of species fruit on insects (Lopez-Quintero et al. 2012). The species fruiting on soil grow well in the wet soils enriched with dead decaying organic matter (dirt) providing nutrients and energy required for growth and development of sporocarps. Sometimes, the soil-inhabiting species dominate the species fruiting on other substrates (Thulasinathan et al. 2018). The soil type, porosity, pH, and supplement accessibility significantly affect the dirt development acting as a substrate for the macrofungal species (Karwa and Rai 2010). The saprotrophic species act as ecosystem scavengers degrading plant debris and animal remains (coprophilous or dung fungi). The rotting efficiency is due the presence of different kinds of enzymes enabling the saprotrophs to degrade a variety of substrates (Conceição et al. 2018). The wood-decaying saprotrophs such as *Armillaria* sp., *Fomes fomentarius,*

Fomitopsis pinicola, *Ganoderma applanatum*, *Hypoxylon* sp., *Lycoperdon perlatum*, *L. pyriforme*, *Pleurotus ostreatus*, *Scutellinia* sp., *Xylaria longipes*, etc. have been found associated with coarse woody debris, stumps, and dead standing trees (Kacprzyk et al. 2014). The wood decomposition rate of wood-deconstructing saprotrophs depends upon wood properties and enzyme activities (Kahl et al. 2017). Krupodorova et al. (2014) reported six types of extracellular enzymes activities in 30 macrofungi fruiting on decaying wood, saprotrophic, entomophilous, and decaying leaf litter and belonging to 6 orders and 18 families. The evolution of mycorrhizae occurred in parallel with the establishment of the first land plant 450–500 million years ago, and this symbiosis still exists in most plant species (Cairney 2000). Of the mycorrhizal fungi, ECM forms present vast diversity and are of great ecological significance. Most of these belong to *Basidiomycota* and to some extent to *Ascomycota* (Smith and Read 2008). These ECM forms provide host plant access to the soil nutrients, e.g., N and P, and improve resistance of host towards drought, disease, and heavy metal excess in exchange for carbon (Nouhra et al. 2005; Toju et al. 2014). Macrofungi in *Agaricomycotina* belong to *Tremellomycetes*, *Dacrymycetes*, and *Agaricomycetes* and exhibit great variation in nutritional mode. The species in *Dacrymycetes* and *Tremellomycetes* are saprotrophic on wood (gelatinous basidiocarp-forming species). *Agaricomycetous* species display diverse nutritional differences and are saprotrophic, plant parasitic, ECM forms, endophytes, mycoparasites, amoebophagous, symbionts, lichens, etc. (Adl et al. 2012). Majority of the *Agaricomycetes* are saprotrophs and are members of all orders, while ECM forms belong to at least 13 orders of the class. Nearly 20,000 to 25,000 ECM fungi form association with 6000 tree species (Rinaldi et al. 2008; Tedersoo et al. 2010). About 90% of plant species are in association with mycorrhizal fungi (Suz et al. 2018). The most common species ECM forms occur in the genera, e.g., *Cantharellus*, *Coltricia*, *Geastrum*, *Laccaria*, *Microporus*, *Paxillus*, *Ramaria*, and *Russula*. Several taxa are common plant parasites, e.g., *Armillaria*, *Coprinellus*, *Fomitiporia*, and *Phellinus* species (Hibbett et al. 2014). The most common mode of nutrition in the members of *Agaricomycetes* is the wood-degrading saprotrophic mode. The wood-degrading saprotrophs in *Agaricomycetes* possess multiple ligninolytic fungal peroxidases (PODs) and many other plant cell wall (PCW) decaying enzymes and hence are capable of producing white rot (resulting in degradation of mainly cellulose and some lignin present in PCW (Floudas et al. 2012; Ruiz-Duenas et al. 2013). The species in the genera such as *Auricularia*, *Pleurotus*, *Stereum*, and *Trichaptum* cause white rot, while some others, e.g., *Fistulina* species, are the common brown rotters (Hibbett et al. 2014). Brown rot (in which mainly cellulose is degraded) in *Polyporales*, *Boletales*, and *Gloeophyllales*, and the evolution of ECM forms, e.g., *Laccaria bicolor* (*Agaricales*), might occur due to the repeated loss of PODs and other PCW decaying enzymes. Some saprotrophic agaricomycetous forms do not cause white rot or brown rot, e.g., *Schizophyllum commune* and *Fistulina hepatica*, and some soil-, litter-, and dung-inhabiting fungi, e.g., *A. bisporus* and *Coprinopsis cinerea* (Ohm et al. 2010; Stajich et al. 2010; Morin et al. 2012). Most of the plant saprotrophic/parasitic or ECM macrofungi occur on the angiosperms and sometimes on gymnosperms, namely, *Abies pindrow*, *Acer cappadocicum, Aesculus indica, Albizia lebbeck, A. procera, Artocarpus chapla-sha, Bambusa vulgaris, Betula utilis, Corylus jacquemontii, Dalbergia sissoo, Juglans regia, Macrophyla mahogoni, Picea smithiana, Pinus wallichiana, Quercus semecarpifolia, Rhododendron arboretum, Tectona grandis*, etc. (Chander 2016; Marzana et al. 2018). Many species have wide host range, while some others dis-play host exclusivity/recurrence (Nogueira-Melo et al. 2017). For example, *A. auricula-judae* shows wide host range infecting *Acacia catechu, Bauhinia varie-gata, Bombax ceiba, Dalbergia sissoo, Ficus benghalensis, Melia azedarach, Tamarindus indica*, and *Toona ciliata*. Some have narrow host range, e.g., *Ganoderma lucidum* gets associated only with *Pyrus pashia, Tamarindus indica, Toona ciliata* and *Ziziphus jujuba*, while *Pleurotus ostreatus* displays host exclusiv-ity infecting only *Melia azedarach* (Pathania and Chander 2018). *Laetiporus sul-phureus*, an edible and wood-decaying fungus, is found associated with bamboo; living and dead oaks; hardwoods such as *Quercus, Prunus, Pyrus, Populus, Salix, Robinia, Fagus, Ceratonia*, and *Eucalyptus*; and occasionally conifers (Verma et al. 2017). The habit of exclusivity/recurrence could be attributed to their mycorrhizal interactions with specific tree species; role in decaying particular types of substrates such as litter, animal dung, etc.; and colonization upon the types of soil formed in a habitat by their interaction with host species as well as rotting of fallen litters (Pradhan et al. 2013). Coprophilous fungi such as species in the genera *Coprinopsis, Conocybe, Lycoperdon, Mycena, Stropharia*, and *Tricholoma* grow on herbivorous animal dung (camel, cattle, buffalo, donkey, goat, horse) rich in the most resistant and undigested remains of feed such as cell wall constituents (cellulose, hemicel-luloses, and lignin), animal gut microbiota, and many other components containing nitrogen (Mohammed et al. 2017; Mumpuni et al. 2020). The coprophilous species contain lytic enzymes and decompose the undigested portion of feed present in the animal dung (Couturier et al. 2016). The macrofungal species form some interesting symbiotic association with insects, e.g., *Termitomyces* species exhibit symbiosis with termites (Nobre et al. 2011). Termitomycetes include tiny *Termitomyces micro-carpus* and giant *T. titanicus* (Tibuhwa et al. 2010). There are 40 species of *Termitomyces* exhibiting symbiosis with termites (Kirk et al. 2001). This type of association is common in Africa and Asia resulting in decay and production of materials needed for termites. About 330 species of the subfamily *Macrotermitinae* are involved in termitomycete cultivation (Mueller et al. 2005). Likewise, 34 spe-cies in the genus *Apterostigma* participate in the cultivation of the coral fungi placed in Pterulaceae (Mehdiabadi and Schultz 2010). Macrofungi exhibit the tendency of multi-trophic life strategies under certain circumstances such as biotrophy and sap-rotrophy, or necrotrophy and saprotrophy, or endophytism and saprotrophy, as these are not obligatory "mutually exclusive." For example, *Armillaria mellea*, a typical necrotrophic basidiomycete of living trees, also shows mycorrhizal relationship with orchids (Kikuchi et al. 2008a, b). Most of the *Pezizomycotina* members are lichen-forming species. About 40% of this group form lichens and four classes, namely, *Arthoniomycetes, Coniocybomycetes, Lichinomycetes*, and *Lecanoromycetes*, consisting of only lichen-forming fungi. Nearly 98% of lichens are *Pezizomycotina* (Grube and Wedin 2016). Other life strategies of *Pezizomycotina*

species are mycorrhizal, plant parasitic, endophytes, animal parasitic and symbionts, mycoparasitic amoebophagous, endolichenic, or endolithic (Stajich et al.
2009; Corsaro et al. 2017; Spatafora et al. 2017). The ascomycetous macrofungi
occur in *Leotiomycetes*, *Pezizomycetes*, and *Sordariomycetes*. Leotiomycetes are
saprotrophs, plant pathogenic, rarely lichen-forming species. Members in the class
Pezizomycetes are saprotrophs, ECM forms, or plant pathogens. The species placed
in the class *Sordariomycetes* can be saprotrophs, plant pathogenic, parasitic on animals and fungi, endophytes, or rarely form lichens (Adl et al. 2012). The representa-
tive ascomycetous macrofungal genera include *Cordyceps*, *Gyromitra*, *Helvella*,
Leotia, *Morchella*, *Peziza*, *Pyronema*, *Terfezia*, and *Tuber*. Entomophagous macrofungi are capable of parasitizing insects and their larvae. *Cordyceps* spp. are known
parasites of an array of insects and their larvae. There are about 500 species of the
genus *Cordyceps* reported worldwide (Sung et al. 2007; Kepler et al. 2012; Dattaraj
et al. 2018).

References

Adl SM, Simpson AG, Lane CE, Luke˘s J, Bass D, Bowser SS, Brown M, Burki F, Dunthorn
    M, Hampl V, Heiss A, Hoppenrath M, Lara E, Le Gall L, Lynn DH, McManus H, Mitchell
    EAD, Mozley-Stanridge SE, Parfrey LW, Pawlowski J, Rueckert S, Shadwick L, Schoch CL,
    Smirnov A, Spiegel FW (2012) The revised classification of eukaryotes. J Eukaryot Microbiol
    59(5):1–45
Alves CR, Urcelay C, da Silveira RMB (2018) Indicator species and community structure of gas-
    teroid fungi (Agaricomycetes, Basidiomycota) in ecosystems of the Atlantic forest in Southern
    Brazil. Braz J Bot 41(3):641–651
Andrew EE, Kinge TR, Tabi EM, Thiobal N, Mih AM (2013) Diversity and distribution of macro-
    fungi (mushrooms) in the Mount Cameroon Region. J Ecol Nat Environ 5(10):318–334
Bajpai A, Rawat S, Johri BN (2019) Fungal Diversity: global perspective and ecosystem dynamics.
    In: Microbial diversity in ecosystem sustainability and biotechnological applications. Springer,
    Singapore, pp 83–113
Borkar P, Doshi A, Navathe S (2015) Mushroom diversity of Konkan region of Maharashtra, India.
    J Threat Taxa 7(10):7625–7640
Brunner I, Frey B, Hartmann M, Zimmermann S, Frank Graf F, Suz LM, Niskanen T, Bidartondo
    MI, Senn-Irlet B (2017) Ecology of alpine macrofungi – combining historical with recent data.
    Front Microbiol 8:2066
Brunner I, Fischer M, RuÈthi J, Stierli B, Frey B (2018) Ability of fungi isolated from plastic
    debris floating in the shoreline of a lake to degrade plastics. PLoS ONE 13(8):1–14
Cairney JWG (2000) Evolution of mycorrhiza systems. Naturwissenschaften 87(11):467–475
Chander H (2016) Diversity and distribution of macrofungi in the Valley of Flowers National Park.
    J Biol Chem Chron 2(2):36–41
Cho SE, Jo JW, Kim NK, Kwag YN, Han SK, Chang KS, Oh SH, Kim CS (2019) Macrofungal
    survey of the Tian Shan mountains, Kyrgyzstan. Mycobiology 47(4):378–390
Cho SE, Kwag YN, Jo JW, Han SK, Oh SH, Kim CS (2020) Macrofungal diversity of urbanized
    areas in southern part of Korea. J Asia Pac Biodivers 13(2):189–197
Conceição AA, Cunha JR, Vieira VO, Pelaéz RD, Mendonça S, Almeida JR, Dias ES, de Almeida
    EG, de Siqueira FG (2018) Bioconversion and biotransformation efficiencies of wild macro-
    fungi. In: Biology of macrofungi. Fungal Biol. Springer, Cham, pp 361–377

Copoţ O, Tănase C (2019) Lignicolous fungi ecology-biotic and abiotic interactions in forest ecosystems. Memoirs of the scientific sections of the Romanian academy 42

Corsaro D, Kohsler M, Wylezich C, Venditti D, Walochnik J, Michel R (2017) New insights from molecular phylogenetics of amoebophagous fungi (Zoopagomycota, Zoopagales). Parasitol Res 17(1):157–167

Couturier M, Tangthirasunun N, Ning X, Brun S, Gautier V, Bennati-Granier C, Silar P, Berrin JG (2016) Plant biomass degrading ability of the coprophilic ascomycete fungus *Podospora anserina*. Biotechnology Adv 34(5):976–983

Dattaraj HR, Jagadish BR, Sridhar KR, Ghate SD (2018) Are the scrub jungles of Southwest India potential habitats of *Cordyceps*? Kavaka 51:20–22

Dorjey K, Kumar S, Sharma YP (2013) Four *Helvella* (Ascomycota: Pezizales: Helvellaceae) species from the Cold Desert of Leh, Ladakh, Jammu and Kashmir, India. J Threat Taxa 5(5):3981–3984

Favre J (1955) Les champignons supérieurs de la zone alpine du Parc National Suisse. Ergebn Wiss, Untersuch. Schweiz 5:1–212

Fernández A, Sánchez S, García P, Sánchez J (2020) Macrofungal diversity in an isolated and fragmented Mediterranean forest ecosystem. Plant Biosyst 154(2):139–148

Floudas D, Binder M, Riley R, Barry K, Blanchette RA, Henrissat B, Martinez AT, Otillar R, Spatafora JW, Yadav JS, Aerts A, Benoit I, Boyd A, Carlson A, Copeland A, Coutinho PM, de Vries RP, Ferreira P, Findley K, Foster B, Gaskell J, Glotzer D, Gorecki P, Heitman J, Hesse C, Hori C, Igarashi K, Jurgens JA, Kallen N, Kersten P, Kohler A, Kues U, Kumar TKA, Kuo A, LaButti K, Larrondo LF, Lindquist E, Ling A, Lombard V, Lucas S, Lundell T, Martin R, McLaughlin DJ, Morgenstern I, Morin E, Murat C, Nagy LG, Nolan M, Ohm RA, Patyshakuliyeva A, Rokas A, Ruiz-Duenas FJ, Sabat G, Salamov A, Samejima M, Schmutz J, Slot JC, St John F, Stenlid J, Sun H, Sun S, Syed K, Tsang A, Wiebenga A, Young D, Pisabarro A, Eastwood DC, Martin F, Cullen D, Grigoriev IV, Hibbett DS (2012) The Paleozoic origin of enzymatic lignin decomposition reconstructed from 31 fungal genomes. Science 336:1715–1719

Ghate SD, Sridhar KR (2016) Contribution to the knowledge on macrofungi in mangroves of the Southwest India. Plant Biosyst 150(5):977–986

Grube M, Wedin M (2016) Lichenized fungi and the evolution of symbiotic organization. Microbiol Spectr 4:749–765

Hawksworth DL (2001) The magnitude of fungal diversity: the 1.5 million species estimate revisited. Microbiol Res 105:1422–1432

Hibbett DS, Bauer R, Binder M, Giachini AJ, Hosaka K, Justo A, Larsson E, Larsson KH, Lawrey JD, Miettinen O, Nagy LG (2014) Agaricomycetes. In: Systematics and evolution. Springer, Berlin/Heidelberg, pp 373–429

Iddison P (2011) Truffles in middle eastern cookery. Emirates Natural History Group, (Patron: H. E. Sheikh Nahayan bin Mubarak Al Nahayan); Available from http://www.enhg.org/alain/phil/truffle/truffle.htm

Jamoni PG (2008) Funghi alpini delle zone alpine superiori e inferiori. Associazione Micologica Bresadola

Jang Y, Jang S, Lee J, Lee H, Lim YW, Kim C, Kim JJ (2016) Diversity of wood-inhabiting polyporoid and corticioid fungi in Odaesan National Park, Korea. Mycobiology 44(4):217–236

Kacprzyk M, Bednarz B, Kuźnik E (2014) Dead trees in beech stands of the Bieszczady National Park: quantitative and qualitative structure of associated macrofungi. Appl Ecol Env Res 12(2):325–344

Kahl T, Arnstadt T, Baber K, Bässler C, Bauhus J, Borken W, Buscot F, Floren A, Heibl C, Hessenmöller D, Hofrichter M (2017) Wood decay rates of 13 temperate tree species in relation to wood properties, enzyme activities and organismic diversities. Forest Ecol Manag 391:86–95

Kark S (2007) Effects of ecotones on biodiversity. In: Levin S (ed) Encyclopedia of biodiversity. Academic Press, San Diego, pp 142–148

Karwa A, Rai M (2010) Tapping into the edible fungi biodiversity of Central India. Biodiversitas 11(2):97–101

Kepler RM, Sung GH, Ban S, Nakagiri A, Chen MJ, Huang B, Li Z, Spatafora JW (2012) New teleomorph combinations in the entomopathogenic genus *Metacordyceps*. Mycologia 104(1):182–197

Kikuchi G, Higuchi M, Morota T, Nagasawa E, Suzuki A (2008a) Fungal symbiont and cultivation test of *Gastrodia elata* Blume (Orchidaceae). J Jpn Bot 83:88–95

Kikuchi G, Higuchi M, Yoshimura H, Morota T, Suzuki A (2008b) In vitro symbiosis between *Gastrodia elata* Blume (Orchidaceae) and *Armillaria* Kummer (Tricholomataceae) species isolated from orchid tuber. J Jpn Bot 83(2):77–87

Kim CS, Jo JW, Kwag YN, Sung GH, Lee SG, Kim SY, Shin CH, Han SK (2015) Mushroom flora of Ulleung-gun and a newly recorded *Bovista* species in the Republic of Korea. Mycobiology 43(3):239–257

Kim CS, Han SK, Nam JW, Jo JW, Kwag Y-N, Han J-G, Sung G-H, Lim YW, Oh S (2017) Fungal communities in a Korean red pine stand, Gwangneung Forest, Korea. J Asia Pac Biodivers 10(4):559–572

Kirk PM, Cannon PF, David JC, Stalpers JA (2001) Ainsworth & Bisby's dictionary of the fungi, 9th edn. CABI Publishing, Wallingford, UK

Krupodorova T, Ivanova T, Barshteyn V (2014) Screening of extracellular enzymatic activity of macrofungi. J Microbiol Biotech Food Sci 3(4):315–318

Lee WD, Lee H, Fong JJ, Oh SY, Park MS, Quan Y, Jung PE, Lim YW (2014) A checklist of the basidiomycetous macrofungi and a record of five new species from Mt. Oseo in Korea. Mycobiology 42(2):132–139

Lopez-Quintero CA, Straatsma G, Franco-Molano AE, Boekhout T (2012) Macrofungal diversity in Colombian Amazon forests varies with regions and regimes of disturbance. Biodivers Conserv 21:2221–2243

Marzana A, Aminuzzaman FM, Chowdhury MSM, Mohsin SM, Das K (2018) diversity and ecology of macrofungi in Rangamati of Chittagong hill tracts under tropical evergreen and semi-evergreen forest of Bangladesh. Advan Res 13(5):1–17

Mehdiabadi NJ, Schultz TR (2010) Natural history and phylogeny of the fungus-farming ants (Hymenoptera: Formicidae: Myrmicinae: Attini). Myrmecol News 13:37–55

Mohammed N, Shinkafi SA, Enagi MY (2017) Isolation of coprophilous mycoflora from different dung types in some local government areas of Niger state, Nigeria. Am J Lif Sci 5(3–1):24–29

Monteiro M, Reino L, Schertler A, Essl F, Figueira R, Ferreira MT, Capinha C (2020) A database of the global distribution of alien macrofungi. Biodivers Data J 8:1–11

Morin E, Kohler A, Baker AR, Foulongne-Oriol M, Lombard V, Nagy LG, Ohm RA, Patyshakuliyeva A, Brun A, Aerts AL, Bailey AM, Billette C, Coutinho PM, Deakin G, Doddapaneni H, Floudas D, Grimwood J, Hilden K, Kues U, Labutti KM, Lapidus A, Lindquist EA, Lucas SM, Murat C, Riley RW, Salamov AA, Schmutz J, Subramanian V, Wosten HA, Xu J, Eastwood DC, Foster GD, Sonnenberg AS, Cullen D, de Vries RP, Lundell T, Hibbett DS, Henrissat B, Burton KS, Kerrigan RW, Challen MP, Grigoriev IV, Martin F (2012) Genome sequence of the button mushroom Agaricus bisporus reveals mechanisms governing adaptation to a humic-rich ecological niche. Proc Natl Acad Sci, USA 109:17501–17506

Mueller UG, Gerardo NM, Aanen DK, Six DL, Schultz TR (2005) The evolution of agriculture in insects. Annu Rev Ecol Evol Syst 36:563–595

Mueller GM, Schmit JP, Leacock PR, Buyck B, Cifuentes J, Desjardin DE, Halling RE, Hjortstam K, Iturriaga T, Larsson KH, Lodge DJ (2007) Global diversity and distribution of macrofungi. Biodiver conserv 16(1):37–48

Mumpuni A, Ekowati N, Wahyono DJ (2020) The existence of coprophilous macrofungi in Banyumas, Central Java, Indonesia. Biodiversitas 21(1):282–289

Naranjo-Ortiz MA, Gabaldón T (2019) Fungal evolution: diversity, taxonomy and phylogeny of the fungi. Biol Rev 94:2101–2137

Nobre T, Koné NA, Konaté S, Linsenmair KE, Aanen DK (2011) Dating the fungus-growing termites' mutualism shows a mixture between ancient codiversification and recent symbiont dispersal across divergent hosts. Mol Ecol (12):2619–2627

Nogueira-Melo GS, Santos PJ, Gibertoni TB (2017) Host-exclusivity and host-recurrence by wood decay fungi (Basidiomycota-Agaricomycetes) in Brazilian mangroves. Acta Bot Bras 31(4):566–570

Nouhra ER, Horton TR, Cazares E, Castellano MA (2005) Morphological and molecular characterization of selected Ramaria mycorrhizae. Mycorrhiza 15(1):55–59

Ohm RA, de Jong JF, Lugones LG, Aerts A, Kothe E, Stajich JE, de Vries RP, Record E, Levasseur A, Baker SE, Bartholomew KA, Coutinho PM, Erdmann S, Fowler TJ, Gathman AC, Lombard V, Henrissat B, Knabe N, Kues U, Lilly WW, Lindquist E, Lucas S, Magnuson JK, Piumi F, Raudaskoski M, Salamov A, Schmutz J, Schwarze FWMR, van Kuyk PA, Horton JS, Grigoriev IV, HAB W (2010) Genome sequence of the model mushroom Schizophyllum commune. Nat Biotechnol 28(9):957–963

Owaid MN (2018) Bioecology and uses of desert truffles (Pezizales) in the Middle East. Walailak J Sci Tech 15(3):179–188

Pathania J, Chander H (2018) Nutritional qualities and host specificity of most common edible macrofungi of Hamirpur district, Himachal Pradesh. J Biol Chem Chron 4(2):86–89

Pradhan P, Dutta AK, Roy A, Bsu SK, Acharya K (2013) Macrofungal diversity and habitat specificity: a case study. Biodivers 14(3):147–161

Priyamvada H, Akila M, Singh RK, Ravikrishna R, Verma RS, Philip L, Marathe RR, Sahu LK, Sudheer P K, Gunthe SS (2017) Terrestrial macrofungal diversity from the tropical dry evergreen biome of southern India and its potential role in aerobiology. PLoS ONE 12:1):1–1)21

Rinaldi AC, Comandini O, Kuyper TW (2008) Ectomycorrhizal fungal diversity: separating the wheat from the chaff. Fungal Divers 33:1–45

Ruiz-Duenas FJ, Lundell T, Floudas D, Nagy L, Barrassa JM, Hibbett D, Martınez AT (2013) Lignindegrading peroxidases in Polyporales: an evolutionary survey based on ten sequenced genomes. Mycologia 105(6):1428–1444

Sarrionandia E, Salcedo I (2018) Macrofungal diversity of holm-oak forests at the northern limit of their distribution range in the Iberian Peninsula. Scand J For Res 33(1):23–31

Semwal KC, Bhatt VK, Stephenson SL (2018) A survey of macrofungal diversity in the Bharsar region, Uttarakhand Himalaya, India. Jasia Pac Biodivers 11(4):560–565

Smith SE, Read DJ (2008) Mycorrhizal symbiosis. Academic Press Ltd, Cambridge, UK

Spatafora JW, Aime MC, Grigoriev IV, Martin F, Stajich JE, Blackwell M (2017) The Fungal tree of life: from molecular systematics to senome-scale phylogenies. Microbiol Spectr 5:3–34

Stajich JE, Berbee ML, Blackwell M, Hibbett DS, James TY, Spatafora JW, Taylor JW (2009) The fungi. Curr Biol 19:840–845

Stajich JE, Wilke SK, Ahren D, Au CH, Birren BW, Borodovsky M, Burns C, Canback B, Casselton LA, Cheng CK, Deng J, Dietrich FS, Fargo DC, Farman ML, Gathman AC, Goldberg J, Guigo R, Hoegger PJ, Hooker JB, Huggins A, James TY, Kamada T, Kilaru S, Kodira C, Kues U, Kupfer D, Kwan HS, Lomsadze A, Li W, Lilly WW, Ma LJ, Mackey AJ, Manning G, Martin F, Muraguchi H, Natvig DO, Palmerini H, Ramesh MA, Rehmeyer CJ, Roe BA, Shenoy N, Stanke M, Ter-Hovhannisyan V, Tunlid A, Velagapudi R, Vision TJ, Zeng Q, Zolan ME, Pukkila PJ (2010) Insights into evolution of multicellular fungi from the assembled chromosomes of the mushroom Coprinopsis cinerea (Coprinus cinereus). Proc Natl Acad Sci USA 107:11889–11189

Sung GH, Hywel-Jones NL, Sung JM, Luangsa-ard JJ, Shrestha B, Spatafora JW (2007) Phylogenetic classification of Cordyceps and the clavicipitaceous fungi. Stud mycol 57:5–59

Suz LM, Sarasan V, Weam JA et al (2018) Positive plant fungal interactions. In: Willis KJ (ed) State of the world fungi. Royal Botanic Gardens, Kew, pp 32–39

Tedersoo L, May TW, Smith ME (2010) Ectomycorrhizal lifestyle in fungi: global diversity, distribution, and evolution of phylogenetic lineages. Mycorrhiza 20(4):217–263

Thiers BM, Halling RE (2018) The macrofungi collection consortium. Appl Plant Sci 6(2):1–7

Thulasinathan B, Kulanthaisamy MR, Nagarajan A, Soorangkattan S, Muthuramalingam JB, Jeyaraman J, Arun A (2018) Studies on the diversity of macrofungus in Kodaikanal region of Western Ghats, Tamil Nadu, India. Biodiversitas 19(6):2283–2293

Tibuhwa DD, Kivaisi AK, Magingo FSS (2010) Utility of the macro-micro morphological characteristics used in classifying the species of *Termitomyces*. Tanz J Sci 36(1):31–45

Toju H, Guimaraes PR, Olesen JM, Thompson JN (2014) Assembly of complex plant–fungus networks. Nat Commun 5(1):1–7

Tripathi NN, Singh P, Vishwakarma P (2017) Biodiversity of macrofungi with special reference to edible forms: a review. J Indian Bot Soc 96(3 and 4):144–187

Verma RK, Asaiya AJK, Chitra C, Vimal P (2017) Diversity of macro-fungi in Central India-IX: *Laetiporus sulphurous*. Van Sangyan 4(11):1–44

Wani AH, Pala SA, Boda RH, Bhat MY (2020) Fungal diversity in the Kashmir Himalaya. In: Biodiversity of the Himalaya: Jammu and Kashmir State. Springer, Singapore, pp 319–341

Chapter 4
Ethnomycology

Ethnomycology is a subject of concern and spotlights the cultural significance and history of uses of macrofungi in human life (Dugan 2011; Brown 2019). Ethnomycological surveys add to our knowledge of various practices involving macrofungi by the locals and are beneficial in a better valorization of their uses (Guissou et al. 2014). Ethnomycological investigations help to decide which species is better to cultivate by conveying information regarding its benefits provided to the locals, unveil the cultural differences among communities with respect to the uses of the species, and also play significant role in the management and conservation plans to save the species involving the locals (Garibay-Orijel et al. 2007). For the past few decades, tribal people are losing their tradition and culture with increase in deforestation, urbanization, and exodus and integration of tribals into urban society (Lachure 2012). Therefore, it becomes important to collect, preserve, and transmit ethnomycological information to the future generations making them aware of the use value of macrofungi and for innovations in the field of mycology. The mycophilic societies/cultures in Greek, Rome, Egypt, China, and Russia collect and use wild fungi as food, in ethnomedicine, and to perform religious rituals (Miles and Chang 2004). In Finland, Italy, Spain, and the United Republic of Tanzania, the opinion towards wild fungi varies even among different parts of the same country (Härkönen et al. 1994; Härkönen 1998). Russians have deep passion for wild fungi evidenced by various sayings. The Estonians describe this Russian passion as "Where there is a mushroom coming up, there is always a Russian waiting for it." In Finnish, Karelia, there is a popular saying "Shouting like Russians in a mushroom forest." Russians are famous for hunting fungi at every weekend (Villarreal and Perez-Moreno 1989; Filipov 1998). In mycophobic countries like Britain, people are afraid of consuming them (Dyke and Newton 1999). However, this fear is disappearing, and the use of macrofungi is expanding because of immigrants from mycophilic regions. In Guatemala, people ignored *B. edulis* but later started consuming it (Flores et al. 2002). Macrofungi always enticed man with their beautiful color and form. People inhabiting different regions of the world consume macrofungi in different ways (Table 4.1). Archaeological records indicate association of edible species of macrofungi with humans living 13,000 years ago in Chile (Rojas and Mansur 1995). The eating of wild macrofungi is reported for the first time from China several hundred years back before the birth of Christ (Aaronson 2000). The use of fungi as human food is evident from the reference of a desert truffle, *Terfezia arenaria*, in the bible as "bread from heaven" and "manna of the Israelites" (Pegler 2002). Edible taxa are known for their delicious taste, exotic aroma, and pleasing flavor. Several species have mythological significance related to their growth and collection. The Gaddang people in the Philippines think that spontaneous lightening promotes fungal growth (Lazo et al. 2015). In West Bengal (India), collection of mushrooms is done on full moon days and new moon days (Dutta and Achariya 2014). The Mexican Indians have mythological thinking that hallucinogenic mushrooms are mediators with God, while Nahum Aztecs considered mushrooms as teonanacatl, meaning God's flesh (Singh 1999). In certain parts of the world, macrofungi have ludic value. The children of different communities in Mexico use sporocarps of *Calvatia* and *Pisolithus* species as balls and projectiles (Hernández-Santiago et al. 2016). Adhikari et al. (2005) conducted an ethnomycological survey in the vicinities of Lumle and Kathmandu Valley of Nepal and gathered information about the culinary uses of 18 macrofungi, medicinal uses of 8 species, and utilization of 3 species for other purposes. In North Central Nigeria, majority of natives consume macrofungi for their nutritional, palatability, and medicinal characteristics (Ayodele et al. 2009). Kimn and Song (2014) recorded 38 species of mushrooms belonging to 33 genera and 22 families (mainly *Tricholomataceae*, *Pleurotaceae*, *Polyporaceae*, and *Hymenochaetaceae*) which are used in 158 types of practices by the tribal communities in Korea. This report shows 24 ways to cook mushrooms such as soups, teas, simmering, roasting, etc. in addition to their medicinal uses. In Gorakhpur Forest Division of Uttar Pradesh (India), the tribals and forest dwellers use *Termitomyces* species as food as well as for medicinal purposes (Srivastava et al. (2011). Fourteen macrofungal genera classified in 11 families and 7 orders have been collected and identified from different habitats (semidesert, gardens, park) associated with decaying roots of dead trees and some under the living trees in Qatar. These macrofungi mainly include truffles, e.g., *Phaeangium lefebvrei*, *Terfezia claveryi*, *Tirmania nivea*, etc. (Al-Thani 2010). Majangir tribe and Wacha inhabitants of Southwestern Ethiopia collect, wash, chop, and use fresh mushrooms in the preparation of stew and soup. People here consider mushrooms as supplement or alternative to meat or fish (Tuno 2001; Teferi et al. 2013). Dutta and Achariya (2014) conducted an ethnomycological survey in West Bengal, India, where the locals and tribals use 34 species of macrofungi of which 31 species are edible and 5 are of medicinal significance, while some of these species solve both the purposes. The inner Baduy people in Cikartawana hamlet, Korea, consume 12 edible macrofungi either fresh or dried in elaborated stew such as soup, steamed, or stir fried (Khastini et al. 2018). In Menge District, Asosa Zone, Benishangul-Gumuz Region of Ethiopia, 20 edible and medicinal mushroom species belonging to 10 genera and 6 families have been identified. Of these, 15 species have been reported to be edible including *Termitomyces schimperi* ranked as first followed by *T. le-testui*,

Table 4.1 Macrofungi with use value to man commonly utilized by different ethnic groups across the globe

Family	Current scientific name	E (edible)/M (medicinal)	Reference
Agaricaceae	Agaricus arvensis, A. bambusicola, A. blazei, A. bisporus, A. campestris, A. crocopeplus, A. semotus, A. subrutilescens, A. subsaharianus, Calvatia cyathiformis, C. lilacina, Chlorophyllum brunneum, C. olivieri, Cyathus limbatus, Gastropila fumosa, Leucocoprinus cepistipes, Lycoperdon lividum, L. marginatum, L. pratense, Macrolepiota mastoidea, M. procera, and Xanthagaricus luteolosporus	E	Valverde et al. (2015), Basumatary and Gogoi (2016), Kinge et al. (2017), Foo et al. (2018), Kamalebo et al. (2018), Liu et al. (2018), Murati and Rexhepi (2018), Robles-García et al. (2018), Rothman (2018), Debnath et al. (2019), Haro-Luna et al. (2019), Kotowski et al. (2019) and Ponce et al. (2019)
	Bovista plumbea, B. pusilla, Calvatia craniiformis, C. gigantea, Chlorophyllum rhacodes, Lepiota cristata, L. perlatum, L. pyriforme, L. subincarnata, and Podaxis pistillaris	M	Smith (1932), Smith (1933), Debnath et al. (2019) and Vishwakarma and Tripathi (2019)
Amanitaceae	Amanita annulatovaginata, A. caesarea, A. chepangiana, A. congolensis, A. craseoderma, A. crassiconus, A. echinulata, A. fulva, A. hemibapha, A. jacksonii, A. laurae, A. marmorata, A. masasiensis, A. novinupta, A. pudica, A. robusta, A. rubescens, A. strobilaceovolvata, A. subviscosa, A. tuza, A. vaginata, and A. xanthogala	E	Montoya et al. (2003), Semwal et al. (2014), Kamalebo et al. (2018), Robles-García et al. (2018), Haro-Luna et al. (2019), Soro et al. (2019), Kotowski et al. (2019) and Ponce et al. (2019)
Auriculariaceae	Auricularia cornea, A. delicata, and A. fuscosuccinea	E	Lazo et al. (2015), Kinge et al. (2017), Foo et al. (2018) and Kamalebo et al. (2018)
	A. auricula-judae and A. nigricans	E and M	Kinge et al. (2017), Foo et al. (2018), Debnath et al. (2019) and Vishwakarma and Tripathi (2019)
Bankeraceae	Sarcodon squamosus	E	Kotowski et al. (2019)
	S. imbricatus	M	Liu et al. (2018)

(continued)

Table 4.1 (continued)

Family	Current scientific name	E (edible)/M (medicinal)	Reference
Boletaceae	*Boletus aestivalis, B. atkinsonii, B. auripes, B. badius, B. loosei, B. speciosus, B. pinophilus, B. regius, B. variipes, Chalciporus luteopurpureus, Fistulinella wolfeana, Harrya chromipes, Imleria badia, Leccinum aurantiacum, L. quercinum, L. rugosiceps, L. pseudoscabrum, L. scabrum, L. variicolor, L. versipelle, Retiboletus griseus, Rubroboletus dupainii, Sutorius luridiformis, Xerocomellus chrysenteron, X. pruinatus, Xerocomus cisalpinus, X. ferrugineus, X. illudens, X. spinulosus,* and *X. subtomentosus*	E	Montoya et al. (2003), Stryamets et al. (2015), Kamalebo et al. (2018), Murati and Rexhepi (2018), Robles-García et al. (2018), Haro-Luna et al. (2019), Kotowski et al. (2019) and Soro et al. (2019)
	B. edulis	E and M	Liu et al. (2018), Debnath et al. (2019) and Haro-Luna et al. (2019)
	Tylopilus felleus	M	Kotowski et al. (2019)
Catathelasmataceae	*Macrocybe titans*	E	Zent et al. (2004)
Coprinaceae	*Coprinus africanus*	E	Soro et al. (2019)
	Coprinus comatus and *Coprinellus disseminatus*	E and M	Kinge et al. (2017), Kamalebo et al. (2018), Rothman (2018) and Vishwakarma and Tripathi (2019)
	Coprinellus micaceus	M	Debnath et al. (2019)
Cordycipitaceae	*Cordyceps militaris* and *C. sinensis*	M	Sung et al. (1998), Semwal et al. (2014) and Debnath et al. (2019)
Cortinariaceae	*Cortinarius caperatus* and *C. mucosus*	E	Kotowski et al. (2019)
Diplocystaceae	*Astraeus hygrometricus*	E and M	Semwal et al. (2014) and Debnath et al. (2019)
Discinaceae	*Gyromitra esculenta*	E and M	Stryamets et al. (2015) and Kotowski et al. (2019)
	Gyromitra sphaerospora	M	Debnath et al. (2019)
Entolomataceae	*Entoloma bloxami* and *Clitopilus passeckerianus*	M	De Mattos-Shipley et al. (2016), Debnath et al. (2019)

(continued)

Table 4.1 (continued)

Family	Current scientific name	E (edible)/M (medicinal)	Reference
Exidiaceae	Pseudohydnum gelatinosum	E	Stoyneva-Gärtner et al. (2017)
Fomitopsidaceae	F. pinicola	M	Vishwakarma and Tripathi (2019)
	Grifola frondosa and L. sulphureus	E and M	Stryamets et al. et al. (2015), Kinge et al. (2017) and Vishwakarma and Tripathi (2019)
Fistulinaceae	Fistulina hepatica	M	Vishwakarma and Tripathi (2019)
Ganodermataceae	Amauroderma omphalodes	E	Zent et al. (2004)
	Ganoderma oerstedii	E and M	Haro-Luna et al. (2019)
	G. applanatum, G. lucidum, and G. tsugae	M	Kamalebo et al. (2018), Debnath et al. (2019) and Vishwakarma and Tripathi (2019)
Geastraceae	Geastrum triplex	M	Debnath et al. (2019)
Gomphaceae	Gomphus clavatus, Ramaria bonii, R. cystidiophora, R. fennica, R. flava, R. flavigelatinosa, R. madagascariensis, R. rubiginosa, R. rubripermanens, R. sanguinea, R. testaceoflava, and R. versatilis	E	Montoya et al. (2003), Kang et al. (2016), Liu et al. (2018) and Larios-Trujillo et al. (2019)
	Gomphus floccosus and Ramaria botrytis	E and M	Semwal et al. (2014), Liu et al. (2018) and Debnath et al. (2019)
	Corallium formosum	M	Debnath et al. (2019)
Gomphidiaceae	Chroogomphus jamaicensis	E	Montoya et al. (2003)
Gyroporaceae	Gyroporus castaneus and G. cyanescens	E	Kotowski et al. (2019)
Helvellaceae	Helvella crispa	E	Montoya et al. (2003) and Robles-García et al. (2018)
	Helvella lacunosa	E and M	Debnath et al. (2019)
	Helvella macropus and H. acetabulum	M	Debnath et al. (2019)
Hericiaceae	Hericium erinaceus	E and M	Semwal et al. (2014) and Liu et al. (2018)
	Hericium coralloides	M	Debnath et al. (2019)

(continued)

Table 4.1 (continued)

Family	Current scientific name	E (edible)/M (medicinal)	Reference
Hydnaceae	*Cantharellus cerinoalbus, C. congolensis, C. densifolius, C. floridulus, C. isabellinus, C. longisporus, C. luteopunctatus, C. miniatescens, C. minor, C. pseudofriesii, C. ruber, C. rufopunctatus, Craterellus tubaeformis, Hydnum ellipsosporum, H. repandum*, and *Pseudocraterellus undulatus*	E	Semwal et al. (2014), Stryamets et al. (2015), Foo et al. (2018), Kamalebo et al. (2018), Kotowski et al. (2019) and Larios-Trujillo et al. (2019)
	Cantharellus cibarius	E and M	Liu et al. (2018), Robles-García et al. (2018), Haro-Luna et al. (2019), Kotowski et al. (2019) and Ponce et al. (2019)
Hydnangiaceae	*Laccaria amethystina, L. bicolor,* and *L. laccata*	E	Montoya et al. (2003), Kotowski et al. (2019) and Larios-Trujillo et al. (2019)
Hygrophoraceae	*Hygrocybe cantharellus, H. miniata, H. chrysodon*, and *H. hypothejus*	E	Montoya et al. (2003), Foo et al. (2018), Kamalebo et al. (2018) and Kotowski et al. (2019)
Hygrophoropsidaceae	*Hygrophoropsis aurantiaca*	E	Larios-Trujillo et al. (2019)
Hymenochaetaceae	*Inonotus hispidus, I. obliquus, Phellinus fastuosus*, and *Ph. linteus*	M	Illana-Esteban (2011), Khastini et al. (2018) and Debnath et al. (2019)
Hypocreaceae	*Hypomyces lactifluorum*	E and M	Robles-García et al. (2018), Haro-Luna et al. (2019), Larios-Trujillo et al. (2019) and Ponce et al. (2019)
Hypoxylaceae	*Daldinia concentrica* and *D. eschscholtzii*	M	Debnath et al. (2019) and Vishwakarma and Tripathi (2019)

(continued)

Table 4.1 (continued)

Family	Current scientific name	E (edible)/M (medicinal)	Reference
Lyophyllaceae	*Calocybe gambosa,* *Termitomyces aurantiacus,* *T. eurrhizus, T. fuliginosus,* *T. le-testui, T. mammiformis,* *T. medius, T. robustus,* *T. schimperi, T. striatus,* *T. tyleranus,* and *T. umkowaani*	E	Montoya et al. (2003), Basumatary and Gogoi (2016), Kinge et al. (2017), Foo et al. (2018), Kamalebo et al. (2018), Rothman (2018), Kotowski et al. (2019) and Soro et al. (2019)
	Lyophyllum decastes and *Termitomyces albuminosus,* *T. clypeatus, T. heimii,* *T. microcarpus,* and *T. titanicus*	E and M	Semwal et al. (2014), Kinge et al. (2017), Liu et al. (2018), Debnath et al. (2019), Larios-Trujillo et al. (2019) and Vishwakarma and Tripathi (2019)
	Hypsizygus tessulatus and *Termitomyces tyleranus*	M	Debnath et al. (2019)
Marasmiaceae	*Calyptella longipes, Marasmius arborescens, M. bekolacongoli,* *M. buzungolo, M. confertus,* *M. oreades,* and *Trogia infundibuliformis*	E	Kamalebo et al. (2018), Robles-García et al. (2018), Haro-Luna et al. (2019) and Kotowski et al. (2019)
Meripilaceae	*Meripilus giganteus*	E	De Leon et al. (2016)
Morchellaceae	*Morchella elata*	E	Kang et al. (2016) and Kotowski et al. (2019)
	M. esculenta	E and M	Semwal et al. (2014), Debnath et al. (2019), Kotowski et al. (2019) and Vishwakarma and Tripathi (2019)
	Disciotis venosa, Morchella angusticeps, M. conica, *M. hybrida, M. vulgaris,* and *Ptychoverpa bohemica*	M	Debnath et al. (2019) and Vishwakarma and Tripathi (2019)
Mycenaceae	*Mycena* sp.	E	Kamalebo et al. (2018)
Omphalotaceae	*Gymnopus dryophilus, G.* sp., and *Lentinula edodes*	E	Foo et al. (2018), Kamalebo et al. (2018) and Robles-García et al. (2018)
Ophiocordycipitaceae	*Ophiocordyceps sinensis*	M	Liu et al. (2018)
Paxillaceae	*Paxillus cuprinus*	E	Kotowski et al. (2019)
Phallaceae	*Phallus indusiatus*	E and M	Liu et al. (2018)
	Phallus impudicus and *P. rubicundus*	M	Debnath et al. (2019)
Pezizaceae	*Peziza repanda*	M	Debnath et al. (2019)
Pyronemataceae	*G. sumneriana* and *Humaria hemisphaerica*	M	

(continued)

Table 4.1 (continued)

Family	Current scientific name	E (edible)/M (medicinal)	Reference
Physalacriaceae	*Armillaria borealis*, *A. lutea*, and *A. mellea*	E	Robles-García et al. (2018), Haro-Luna et al. (2019) and Kotowski et al. (2019)
	Flammulina velutipes	E and M	Smiderle et al. (2006) and Debnath et al. (2019)
Pleurotaceae	*Pleurotus cornucopiae*, *P. cystidiosus*, *P. djamor* var. *djamor*, *P. djamor* var. *roseus*, *P. eryngii*, *P. giganteus*, *P. sapidus*, and *P. opuntiae*	E	Semwal et al. (2014), Valverde et al. (2015), Parveen et al. (2017), Kinge et al. (2017), Foo et al. (2018), Kamalebo et al. (2018), Haro-Luna et al. (2019) and Kotowski et al. (2019)
	Pleurotus ostreatus and *Lentinus tuber-regium*	E and M	Valverde et al. (2015), Kotowski et al. (2019), Vishwakarma and Tripathi (2019) and Foo et al. (2018)
Pluteaceae	*Volvariella bombycina* and *V. earlei*	E	Haro-Luna et al. (2019) and Soro et al. (2019)
	Volvariella volvacea	E and M	Basumatary and Gogoi (2016), Foo et al. (2018) and Vishwakarma and Tripathi (2019)
	Volvariella glandiformis and *V. indica*	M	Vishwakarma and Tripathi (2019)
Polyporaceae	*Coriolopsis caperata, Daedalea flavida, Favolus tenuiculus, Lentinus brunneofloccosus, L. crinitus, L. levis, L. polychrous, L. sajor-caju, Lentinus squarrosulus, L. strigosus, Lignosus* sp., *L. sulphureus*, and *Polyporus umbellatus*	E	Zent et al. (2004), Stryamets et al. (2015), Basumatary and Gogoi (2016), Parveen et al. (2017), Kamalebo et al. (2018), Foo et al. (2018) and Haro-Luna et al. (2019)
	Echinochaete brachypora and *Lentinus tuber-regium*	E and M	Foo et al. (2018) and Soro et al. (2019)
	F. fomentarius, Lentinus tigrinus, Lenzites betulina, Microporus xanthopus, Polyporus squamosus, Pycnoporus cinnabarinus, P. sanguineus, Trametes gibbosa, T. hirsuta, and *T. versicolor*	M	Parveen et al. (2017), Kamalebo et al. (2018), Debnath et al. (2019), Haro-Luna et al. (2019) and Vishwakarma and Tripathi (2019)

(continued)

Table 4.1 (continued)

Family	Current scientific name	E (edible)/M (medicinal)	Reference
Psathyrellaceae	*Psathyrella tuberculata*	E and M	Soro et al. (2019)
	Coprinopsis atramentaria and *Panaeolus papilionaceus*	M	Debnath et al. (2019) and Vishwakarma and Tripathi (2019)
Rhizopogonaceae	*Rhizopogon roseolus* and *R. villosulus*	M	Debnath et al. (2019)
Rikenellaceae	*Cotylidia aurantiaca*	E	Kamalebo et al. (2018)
Russulaceae	*Lactarius acutus, L. azonites, L. camphoratus, L. deliciosus, L. deterrimus, L. indigo, L. pubescens, L. resimus, L. rimosellus, L. rubrilacteus, L. salmonicolor, L. saponaceus, L. tenellus, L. torminosus, L. vellereus, L. volemus, Lactifluus annulatoangustifolius, L. flammans, L. gymnocarpoides, L. heimii, L. luteopus, L. pelliculatus, L. volemoides, Russula aeruginea, R. alutacea, R. annulata, R. aurea, R brevipes, R. cellulata, R. claroflava, R. ciliata, R. congoana, R. cyanoxantha, R. delica, R. grisea, R. inflata, R. integra, R. lepida, R. nitida, R. rosea, R. meleagris, R. nigricans, R. ochroleuca, R. oleifera, R. porphyrocephala, R. pruinata, R. sesemoindu, sesenagula, R. striatoviridis*, and *R. virescens*	E	Montoya et al. (2003), Comandini et al. (2012), Semwal et al. (2014), Stryamets et al. (2015), Kang et al. (2016), Kamalebo et al. (2018), Liu et al. (2018), Robles-García et al. (2018), Haro-Luna et al. (2019), Kotowski et al. (2019), Ponce et al. (2019) and Soro et al. (2019)
	Russula aquosa and *R. emetica*	M	Vishwakarma and Tripathi (2019)
Sarcoscyphaceae	*Cookeina speciosa, C. sulcipes*, and *C. tricholoma*	E	Foo et al. (2018) and Kamalebo et al. (2018)
Sarcosomataceae	*Galiella rufa*	E	Foo et al. (2018)
Schizophyllaceae	*S. commune*	E and M	Kamalebo et al. (2018), Foo et al. (2018), Debnath et al. (2019) and Vishwakarma and Tripathi (2019)
Sclerodermataceae	*Scleroderma citrinum* and *Calostoma insigne*	E	De Leon et al. (2016), Foo et al. (2018) and Kotowski et al. (2019)
	Pisolithus arrhizus	M	Debnath et al. (2019)

(continued)

Table 4.1 (continued)

Family	Current scientific name	E (edible)/M (medicinal)	Reference
Sparassidaceae	*S. crispa*	E and M	Semwal et al. (2014), Kotowski et al. (2019) and Vishwakarma and Tripathi (2019)
	Sparassis spathulata	M	Debnath et al. (2019)
Stereaceae	*Stereum hirsutum*	M	Vishwakarma and Tripathi (2019)
Strophariaceae	*Gymnopilus zenkeri, Hebeloma mesophaeum,* and *Protostropharia semiglobata*	E	Montoya et al. (2003), Basumatary and Gogoi (2016) and Kamalebo et al. (2018)
	Stropharia rugosoannulata	E and M	Alinia-Ahandani et al. (2018) and Liu et al. (2018)
Suillaceae	*Suillus bovinus, S. granulatus, S. grevillei, S. luteus, S. pseudobrevipes, S. salmonicolor,* and *S. variegatus*	E	Montoya et al. (2003), Stryamets et al. (2015), Robles-García et al. (2018), Kotowski et al. (2019) and Ponce et al. (2019)
Tapinellaceae	*Tapinella panuoides*	E	Kamalebo et al. (2018)
Tremellaceae	*Tremella fuciformis* and *T. mesenterica*	E	Stoyneva-Gärtner et al. (2017) and Foo et al. (2018)
Tricholomataceae	*Clitocybe fragrans, C. squamulosa, Collybia aurea, Gerhardtia cibaria, Infundibulicybe gibba, Lepista nuda, L. sordida, Leucopaxillus giganteus, Tricholoma equestre,* and *T. portentosum*	E	Montoya et al. (2003), Trutmann et al. (2012), Kang et al. (2016), Kamalebo et al. (2018), Rothman (2018), Robles-García et al. (2018), Kotowski et al. (2019) and Ponce et al. (2019)
	Lycoperdon perlatum and *Tricholoma matsutake*	E and M	Valverde et al. (2015), Liu et al. (2018), Robles-García et al. (2018), Rothman (2018) and Vishwakarma and Tripathi (2019)
Tuberaceae	*Tuber aestivum, T. brumale, T. nigrum,* and *T.* sp.	E	Murati and Rexhepi (2018), Debnath et al. (2019) and Kotowski et al. (2019)
Xylariaceae	*Xylaria hypoxylon* and *X. polymorpha*	M	Debnath et al. (2019) and Vishwakarma and Tripathi (2019)

T. microcarpus, and *T. eurhizus* (Sitotaw et al. 2020). In India, the edible and medicinal practices of mushrooms are quite common dating back to 1700–1100 BC (Wasson 1971). Edible mushrooms have been used in different states of India such as Odisha, West Bengal, Assam, Manipur, Nagaland, and Arunachal Pradesh (Baruah et al. 1971; Sing and Sing 1993; Sing et al. 2002; Sarma et al. 2010; Tanti et al. 2011). The most commonly consumed species in these regions belong to the genera *Agaricus, Auricularia, Boletus, Cantharellus, Coprinus, Lactarius, Lentinus, Lycoperdon, Morchella, Ramaria, Russula, Schizophyllum, Termitomyces, Tricholoma*, etc. (Choudhary et al. 2015). An ethnomycological survey revealed the use of 12 edible macrofungal species, namely, *Agaricus campestris, Helvella compressa, Morchella conica, M. esculenta, M. deliciosa, Ramaria botrytis, Lactarius deliciosus, Rhizopogon vulgaris, Sparassis crispa, Gyromitra* sp., *Hygrophorus* sp., and *Lycoperdon* sp., which are cooked and eaten as food by the local tribal people in Kinnaur District of Himachal Pradesh (India). Out of these, *Morchella* species are highly prized and traded in the markets between Rs.8,000 and 12,000/kg (Chauhan et al. 2014).

The collection and marketing of macrofungal species in local markets proves as a good income source for the locals and tribal people. In Nepal, 49 taxa of macrofungi including *L. sulphureus, B. edulis, C. cibarius, P. ostreatus, A. hygrometricus*, and *T. heimii* are used as food and in ethnomedicine (Pandey et al. 2006). Some of these species are being sold fresh in the local markets, while few are sold in dried forms, e.g., *Ramaria* spp. at NPR 100/100 g and dried *M. conica* at NPR 3000 to NPR 7000/kg (Kharel and Rajbhandary 2005). In Mexico, 30 edible mushrooms are known being collected for personal use and sold in the market by the locals. People in the State of Tlaxcala fetch 15.44 Mexican pesos (2.54 USD)/day from mushroom selling in 1995. The average earnings for mushroom collecting per family in Javier Mina range from 62 to 82 Mexican pesos (10.27–13.47 USD)/day (Montoya et al. 2008). *Lactarius, Russula, Cantharellus*, and *Amanita* species are commonly collected and consumed among Benna and Hehe communities inhabiting southern highlands of Tanzania. These people earn 500–650 USD per collection season from 1000 to 1500 kg mushrooms, while retailers make 750–1000 USD per season from 750 at 800 kg of mushrooms (Chelela et al. 2014). Cortés et al. (2018) reported 21 wild macromycete species, of which *R. chloroides, Ampulloclitocybe clavipes, A. mellea, B. edulis*, and *L. indigo* have been found commercially important. In India, local people of Jammu Province collect, consume *Geopora arenicola* widely, and also sell it in the local markets at INR 50/kg. Other species sold here include *Ramaria formosa* at INR 40–50/kg, *Ramaria flavobrunnescens* at INR 30/kg, *S. crispa* at Rs. 50–60/kg, and *T. striatus* at INR 70/kg. Besides these, *Morchella* spp. are sold by the villagers to the middleman at INR 2000–3000/kg, who further sell these to the wholesalers at INR 5000–7000/kg (Kumar and Sharma 2010). The collection of wild macrofungi and selling them in local markets are reported to be a common tradition among Khasi tribe of Meghalaya, an eastern state of India (Khaund and Joshi 2013). The edible species commonly traded here are *G. floccosus* at INR 200/kg, *Tricholoma viridiolivaceum* at INR 200/kg, *Tricholoma saponaceum* at INR 200/kg, *Craterellus lateritius* at INR 280/kg, *L. volemus* and *C. cibarius* at INR 280/

kg, *Laccaria fraterina* at INR 280–300/kg, *Albatrellus* sp. at INR 200/kg, *Ramaria* sp. at INR 250–300/kg, and *Clavulina* sp. at INR 300–350/kg. Wild edible and medicinal truffles (*Pezizales*) such as species in the genera *Terfezia*, *Tirmania*, *Tuber*, *Picoa*, and *Phaeangium* have feeding, medicinal, and economic significance and are also sold in the local markets (Owaid 2018). Several edible species are considered medicinal or to be a medicinal food that helps in curing a number of ailments being rich in an array of nutrients (Bautista-González and Moreno-Fuentes 2014). There are approximately 7000 species of mushrooms which are considered to have varying levels of edibility. More than 700 species are documented to have considerable pharmacological properties (Jo et al. 2014). Medicinal species are used either directly as sporocarps or in the form of powder/paste for the treatment of different diseases. In communities of Baduy tribe, Indonesia, *Amauroderma* sp. is considered as a good analgesic, which helps to cure fever and immunity deficiency; *G. lucidum* and *G. applanatum* act against cancer and immunodeficiency disorders; *F. fomentarius* is good in fever; and *Ph. linteus* acts against cancer and is known to improve blood circulation (Khastini et al. 2018). An ethnomycological survey in Chuxiong City, Yunnan (China), reveals the medicinal uses of *L. volemus* (acting against cancer, lowering blood pressure, improving working of the liver, detoxifying, and useful in diuresis), *O. sinensis* (working against cancer, nourishing lungs, and regulating breath), *Ramaria botrytis* (good for the stomach and acting against cancer), *P. indusiatus* (good for the lungs, regulating breath, improving sleep, and stimulating immunity), *S. crispa* (as antiaging, declining cholesterol, anticancer, and immunostimulatory agent), *Stropharia rugosoannulata* (nourishing brain), *T. matsutake* (boosting immunity, acting against aging and cancer, elevating cardiovascular health, good for gastrointestinal functions, and protecting liver), and *T. albuminosus* (as immunostimulant and sleep enhancer) (Liu et al. 2018). As per the survey of Apetorgbor et al. (2006), *A. auricula-judae* and *Trametes* sp. are good blood tonic; *Collybia* sp. works against fever and as blood tonic; *C. disseminatus* and *S. commune* are good blood tonic and help to treat eye infection; *Lentinus tuber-regium* acts against asthma; *G. lucidum* helps to cure cancer; *T. clypeatus* is used to treat rheumatism and diarrhea; *T. globules* is a blood tonic and helps in lowering high blood pressure; *T. microcarpus* is a blood tonic and lowers blood glucose; *T. robustus* is a blood tonic and *V. volvacea* lowers high blood pressure; *C. cibarius* cleanses the liver, increases vision, nourishes lungs, regulates breath, and is good in diuresis; *H. crispa* helps in curing pernicious anemia, improves neurasthenia, and lowers cholesterol; *H. erinaceus* exhibits antiaging, anti-inflammatory, antitumor, and antiulcer characteristics, protects the liver, boosts immunity, and enhances blood circulation; and *L. deliciosus* and *Sarcodon imbricatus* possess anticancer properties, provide strength to the body, condition the stomach, are analgesic, nourish lungs, and manage breath in parts of Ghana. Okigbo and Nwatu (2015) documented the use of six fungal species by the people of Anambra State (Nigeria), viz., *A. auricula-judae* and *L. squarrosulus* to treat infertility and anemia; *D. concentrica* to cure stomach upset; *Termitomyces* sp. in anemia, weakness, and high blood pressure; *Trichobatrachus robustus* in anemia and high blood pressure; and *V. volvacea* only in anemia. Ayta communities in Bataan, Philippines, use macrofungi as remedy against weakness, cough, common cold, and poor eyesight (Tantengco and Ragragio 2018). In Cameroon, *Auricularia nigricans* is used during pregnancy and against nausea, *D. concentrica* to treat hypertension, *G. applanatum* and *T. versicolor* to boost the immune system, *L. squarrosulus* in system cleaning, *Polyporus dictyopus* against abdominal pain and headaches, *T. microcarpus* in strengthening bones in children and against fever, *Lycoperdon pratense* against fever, and *Xylaria* sp. against hypertension and fever (Teke et al. 2018). In Chakrata, Dehradun, a hilly region of Northwest India, *A. hygrometricus* is used against ear puss; *A. auricula-judae* enhances milk secretion; the spore mass of *Geastrum triplex*, *Lycoperdon perlatum*, and *P. pistillaris* is used as burn remedy; *Phallus* sp. is an aphrodisiac; *Stereum* sp. acts as a wound healer and truffle; and *Tuber* sp. is used as a mouth freshener (Kumar et al. 2017). *Agaricus augustus* is a food and general health tonic, a supplementary diet to patients suffering from asthma, stroke, heart ailments, and diabetes. *A. bisporus* acts as a general health tonic, immunomodulator, and antidiabetic agent. *B. edulis* is useful after deliveries taken alone or with dandelion as expectorant, is antidepressant, and is used to treat lumbago, leg pains, numbness in limbs and tendon discomfort, against respiratory tract infections, and frost bites. *C. cibarius* is a natural health tonic, heals wounds, and cures bone ailments, general weakness, respiratory ailments, and diabetes. *Disciotis venosa* cures common cold and is used as a tonic and immunomodulatory agent. *F. velutipes* treats skin ailments and provides cure to hypertensive patients. *F. fomentarius* strengthens the immune system and works against cough and common cold. *G. sumneriana* is a stimulant, tonic, anticold agent, and immunomodulator. *H. acetabulum* cures chronic cough. *H. crispa* is a brain tonic; provides cure in asthma, cold, cough, and diabetes; and releases intestinal inflammation and mouth ulceration. *H. coralloides* acts as a sex stimulant and cures gastric irritation and heart burns. *L. deliciosus* is taken as food; is a brain tonic; works against anger, dry cough, and asthma; and promotes digestion. *M. conica* treats heart ailments, is a brain tonic and sex stimulant, and cures arthritis and general weakness. *M. hybrida* is a brain tonic and cures heart ailments, body weakness, and arthritis. *M. vulgaris* heals wounds. *P. impudicus* is a tonic against weakness. *Ramaria stricta* helps to treat asthma and other respiratory ailments, is blood purifier, enhances skin color, and cures eye ailments. *R. villosulus* works against kidney stones and urinary tract infections and cures fatty liver and asthma. *R. roseolus* is used to treat urinary tract infections. *S. crispa* is a general tonic, blood purifier, and anticold agent and treats memory loss, depression, anger, and chest pain. *S. spathulata* is a health tonic, anticold agent, and blood purifier; cures skin diseases such as rashes, itching, and dryness; and heals wounds. *L. perlatum* and *L. pyriforme* are used to heal wounds and frost bites in northern districts of Jammu and Kashmir, India (Malik et al. 2017). In Tamil Nadu (India), *D. concentrica* is used against skin irritation and helps in healing wounds, *C. gigantea* helps to cure upset stomach and stomachache in woman during menstruation, and *P. pistillaris* cures skin diseases and skin burns (Thangaraj et al. 2017). There is a long list of macrofungi with use value in human health which keeps on extending with the continuous discovery of new taxa. However, the ethnomycological knowledge available on many taxa is not yet enough which requires a thorough and extensive exploration.

References

Aaronson S (2000) Fungi. In: The Cambridge world history of food. Cambridge University Press, Cambridge, UK, pp 313–336

Adhikari MK, Devkota S, Tiwari RD (2005) Ethnomycological knowledge on uses of wild mushrooms in Western and Central Nepal. Our Nature 4:13–19

Alinia-Ahandani E, Fazilati M, Alizadeh Z, Ani Boghozian A (2018) The introduction of some mushrooms as an effective source of medicines in Iran Northern. Biol Med 10(5):1–5

Al-Thani RF (2010) Survey of macrofungi (including truffles) in Qatar. KBM J Biol 1(2):26–29

Apetorgbor MM, Apetorgbor AK, Obodai M (2006) Indigenous knowledge and utilization of edible mushrooms in parts of Southern Ghana. Ghana J Forestry 19(1):20–34

Ayodele SM, Akpaja EO, Adamu Y (2009) Some edible and medicinal mushrooms found in Igala land in Nigeria and their sociocultural and ethnomycological uses. In: Proceedings of the 5th International medicinal mushroom conference, Nantong, China, pp 526–531

Baruah HK, Sing DK, Islam M (1971) On the distribution of higher basidiomycetes in the Sibsagar district, India: Assam. Bull Bot Surv 13(3–4):285–289

Basumatary M, Gogoi M (2016) Uses of wild edible macrofungi by Bodo community of Kokrajhar district, Assam, India. Trop Plant Res 3(1):176–181

Bautista-González J, Moreno-Fuentes A (2014) Los hongos medicinales en México. In: Moreno-Fuentes A, Garibay-Orijel R (eds) La etnomicología en México, estado del arte. CONACYT, UAEH, and UNAM, Mexico, pp 145–176

Brown M (2019) Yi Ethnomycology: wild mushroom knowledge and use in Yunnan, China. J Ethnobiol 39(1):131–157

Chauhan RS, Tiwari D, Bisht AS, Shukla A (2014) Ex situ conservation of medicinal and aromatic plants in Bharsar, Uttarakhand, India. Med Plants 6(4):282–292

Chelela L, Chacha M, Matemu A (2014) Wild edible mushroom value chain for improved livelihoods in southern highlands of Tanzania. Am J Res Commun 2(8):1–14

Choudhary M, Devi R, Datta A, Kumar A, Jat HS (2015) Diversity of wild edible mushrooms in Indian subcontinent and its neighboring countries. Recent Adv Biol Med 1:69–76

Comandini O, Erős-Honti Z, Jakucs E, Arzú RF, Leonardi M, Rinaldi AC (2012) Molecular and morpho-anatomical description of mycorrhizas of *Lactarius rimosellus* on *Quercus* sp. with ethnomycological notes on *Lactarius* in Guatemala. Mycorrhiza 22(4):279–287

Cortés LEUC, García AV, Ruan-Soto F (2018) Ethnomycology and mushroom selling in a market from Northwest Puebla, México. Scientia Fungorum 47:47–55

De Leon AM, Kalaw SP, Dulay RM, Undan JR, Alfonzo DO, Undan JQ, Reyes RG (2016) Ethnomycological survey of the Kalanguya indigenous community in Caranglan, Nueva Ecija, Philippines. Curr Res Env Appl Mycol 6(2):61–66

De Mattos-Shipley KMJ, Ford KL, Alberti F, Banks AM, Bailey AM, Foster GD (2016) The good, the bad and the tasty: the many roles of mushrooms. Stud Mycol 85:125–157

Debnath S, Debnath B, Das P, Saha AK (2019) Review on an ethnomedicinal practices of wild mushrooms by the local tribes of India. J Appl Pharm Sci 9(8):144–156

Dugan FM (2011) Conspectus of world ethnomycology: fungi in ceremonies, crafts, diets, medicines and myths. American Phytopathological Society, US

Dutta AK, Achariya K (2014) Traditional and ethno-medicinal knowledge of mushrooms in West Bengal, India. Asian J Pharm Clin Res 7(4):161–164

Dyke AJ, Newton AC (1999) Commercial harvesting of wild mushrooms in Scottish forests: is it sustainable? Scott Forest 53:77–85

Filipov D (1998) Mushroom season has Russians in fungi frenzy. Boston Globe

Flores R, Bran MDC, Honrubia M (2002) Edible mycorrhizal mushrooms of the West Highland Guatemala. In: Hall IR, Wang Y, Zambonelli A, Danell E (eds) Edible ectomycorrhizal mushrooms and their cultivation. Proceedings of the second international conference on edible mycorrhizal mushrooms. July 2001, Christchurch. CD-ROM. New Zealand Institute for Crop and Food Research Limited, Christchurch

Foo SF, Saikim FH, Kulip J, Seelan JSS (2018) Distribution and ethnomycological knowledge of wild edible mushrooms in Sabah (northern Borneo), Malaysia. J Trop Biol Conserv 15:203–222

Garibay-Orijel R, Caballero J, Estrada-Torres A, Cifuentes J (2007) Understanding cultural significance, the edible mushrooms case. J Ethnobiol Ethnomed 3:1):1–1):4

Guissou ML, Guelly AK, Lamèga D (2014) Biodiversity and sustainable use of wild edible fungi in the Sudanian centre of endemism: A plea for valorisation. Ectomycorrhizal symbioses in tropical and neotropical forests 241

Härkönen M (1998) Uses of mushrooms by Finns and Karelians. Int J Circumpolar Health 57(1):40–55

Härkönen M, Saarimäki T, Mwasumbi L (1994) Edible and poisonous mushrooms of Tanzania. African J Mycol Biotechnol 2(2):99–123

Haro-Luna MX, Ruan-Soto F, Guzmán-Dávalos L (2019) Traditional knowledge, uses, and perceptions of mushrooms among the Wixaritari and mestizos of Villa Guerrero, Jalisco, Mexico. IMA Fungus 10:16):1–16)14

Hernández-Santiago F, Pérez-Moreno J, Xoconostle-Cázares B, Almaraz-Suárez JJ, Ojeda-Trejo E, Mata G, Díaz-Aguilar I (2016) Traditional knowledge and use of wild mushrooms by Mixtecs or Ñuu savi, the people of the rain, from Southeastern, Mexico. J Ethnobiol Ethnomed 12:13–35

Illana-Esteban C (2011) Medicinal interest of "Chaga" (*Inonotus obliquus*). Bol Soc Micol Madrid 35:175–185

Jo WS, Hossain MA, Park SC (2014) Toxicological profiles of poisonous, edible and medicinal mushrooms. Mycobiology 42(3):215–220

Kamalebo HM, Malale HNSW, Ndabaga CM, Degreef J, De Kesel A (2018) Uses and importance of wild fungi: traditional knowledge from the Tshopo province in the Democratic Republic of the Congo. J Ethnobiol Ethnomed 14(13):1–13

Kang J, Kang Y, Ji X, Guo Q, Jacques G, Pietras M, Łuczaj N, Li D, Łuczaj L (2016) Wild food plants and fungi used in the mycophilous Tibetan community of Zhagana (Tewo County, Gansu, China). J Ethnobiol Ethnomed 12:21):1–21)13

Kharel S, Rajbhandary S (2005) Ethnomycological knowledge of some wild mushrooms in Bhardeo, Lalitpur, Nepal. Nepal Jour Pl Sci 1:45–49

Khastini RO, Wahyuni I, Saraswati I (2018) Ethnomycology of bracket fungi in Baduy tribe, Indonesia. Biosaintifika 10(2):424–432

Khaund P, Joshi SR (2013) Wild edible macrofungal species consumed by the Khasi tribe of Meghalaya, India. Indian J Nat Prod Resour 4(2):197–204

Kimn H, Song MJ (2014) Analysis of traditional knowledge for wild edible mushrooms consumed by residents living in Jirisan National Park (Korea). J Ethnopharmacol 153(1):90–97

Kinge TR, Apalah NA, Nji TM, Acha AN, Mih AM (2017) Species richness and traditional knowledge of macrofungi (mushrooms) in the Awing forest reserve and communities, Northwest region, Cameroon. J Mycol 2017(2809239):1–9

Kotowski MA, Pietras M, Łuczaj L (2019) Extreme levels of mycophilia documented in Mazovia, a region of Poland. J Ethnobiol Ethnomed 15(12):1–19

Kumar S, Sharma YP (2010) Pezizales from Jammu and Kashmir-1. Indian J Mushrooms 28:23–32

Kumar M, Harsh NSK, Prasad R, Pandey VV (2017) An ethnomycological survey of Jaunsar, Chakrata, Dehradun, India. JoTT 9(9):10717–10725

Lachure PS (2012) Exploration of some medicinal plants used by tribals from Digras region of district Yavatmal, Maharashtra, India. Int J Sci Res Publ 2:3):1–3):4

Larios-trujillo C, Ruan-soto F, Herrerías-diego Y, Blanco-garcía A (2019) local knowledge and economical significance of commercialized wild edible mushrooms in the markets of Uruapan, Michoacan, Mexico. Econ Bot 73(2):200–216

Lazo CRM, Kalaw SP, De Leon AM (2015) Ethnomycological survey of macrofungi utilized by Gaddang communities in Nueva Vizcaya, Philippines. Curr Res Envn Appl Mycol 5(3):256–262

Liu D, Cheng H, Bussmann RW, Guo Z, Liu B, Long C (2018) An ethnobotanical survey of edible fungi in Chuxiong city, Yunnan, China. J Ethnobiol Ethnomed 14(42):1–10

Malik AR, Wani AH, Bhat MY, Parveen S (2017) Ethnomycological knowledge of some wild mushrooms of northern districts of Jammu and Kashmir, India. Asian J Pharm Clin Res 10(9):399–405

Miles PG, Chang ST (2004) Mushrooms: cultivation, nutritional value, medicinal effect and environmental impact. CRC Press

Montoya A, Hernandez-Totomoch O, Estrada-Torres A, Kong A, Caballero J (2003) Traditional knowledge about mushrooms in a Nahua community in the state of Tlaxcala, Mexico. Mycologia 95(5):793–806

Montoya A, Hernández N, Mapes C, Kong A, Estrada-Torres A (2008) The collection and sale of wild mushrooms in a community of Tlaxcala, Mexico. Econ Bot 62:413–424

Murati E, Rexhepi B (2018) Edible and poisonous mushrooms. ECOTEC-J Sci Env Technol 1(1):41–44

Okigbo RN, Nwatu CM (2015) Ethnostudy and usage of edible and medicinal mushrooms in some parts of Anambra state, Nigeria. Nat Resour 6(1):79–89

Owaid MN (2018) Bioecology and uses of desert truffles (Pezizales) in the Middle East. Walailak J Sci Tech 15(3):179–188

Pandey N, Devkota S, Christensen M, Budathoki U (2006) Use of wild mushrooms among the Tamangs of Nepal. Nepal J Sci Technol 7:97–104

Parveen A, Khataniar L, Goswami G, Hazarika DJ, Das P, Gautom T, Barooah M, Boro RC (2017) A study on the diversity and habitat specificity of macrofungi of Assam, India. Int J Curr Microbiol App Sci 6(12):275–297

Pegler DN (2002) Useful fungi of the world: the 'Poor man's tuffles of Arabia' and 'Manna of the Israelites'. Mycologist 16(1):8–9

Ponce JPM, Hernández Calderón MA, Comandini O, Rinaldi AC, Arzú RF (2019) Ethnomycological knowledge among Kaqchikel, indigenous Maya people of Guatemalan highlands. J Ethnobiol Ethnomed 15(36):1–24

Robles-García D, Suzán-Azpiri H, Montoya-Esquivel A, García-Jiménez J, Esquivel-Naranjo EU, Yahia E, Landeros-Jaime F (2018) Ethnomycological knowledge in three communities in Amealco, Querétaro, México. J Ethnobiol Ethnomed 14(7):1–13

Rojas C, Mansur E (1995) Ecuador: informaciones generales sobre productos non madereros en Ecuador. In: Memoria, consulta de expertos sobre productos forestales no madereros para America Latina y el Caribe, pp 208–223

Rothman S (2018) Edible macrofungi of Namibia's thorn bush savanna bioregion and their potential for sustainable development. Master's thesis. Natural Resource Management, Raseborg

Sarma TC, Sarma I, Patiri BN (2010) Wild edible mushrooms used by some ethnic tribes of Western Assam. Bioscan 10(3):613–625

Semwal KC, Stephenson SL, Bhatt VK, Bhatt RP (2014) Edible mushrooms of the North Western Himalaya, India: a study of indigenous knowledge, distribution and diversity. Mycosphere 5(3):440–461

Sing NI, Sing SM (1993) Wild Edible fleshy fungal flora of Manipur. Bioveel 4(2):153–158

Sing NI, Sing SM, Th C (2002) Fleshy fungi of Manipur. In: Plant genetic diversity: exploration, evaluation, conservation. Affiliated East West Press Pvt Ltd, New Delhi, pp 9–13

Singh J (1999) Ethnomycology and folk remedies: fact and fiction. In: From ethnomycology to fungal biotechnology. Springer, Boston, pp 11–17

Sitotaw R, Lulekal E, Abate D (2020) Ethnomycological study of edible and medicinal mushrooms in Menge district, Assosa zone, Benishangul-Gumuz region, Ethiopia. J Ethnobiol Ethnomed 16:11):1–11)14

Smiderle FR, Carbonero ER, Mellinger CG, Sassaki GL, Gorin PAJ, Iacomini M (2006) Structural characterization of a polysaccharide and a α-glucan isolated from the edible mushroom *Flammulina velutipes*. Phytochem 67(19):2189–2196

Smith (1932) Ethnobotany of the Ojibwe Indians. Bull Public Mus Milwaukee 4(3):327–524

Smith (1933) Ethnobotany of the forest Potawatomi Indians. Bull Public Mus Milwaukee 7(1):1–230

Soro B, Koné NA, Vanié-Léabo LPL, Konaté S, Bakayoko A, Koné D (2019) Phytogeographical and sociolinguistical patterns of the diversity, distribution, and uses of wild mushrooms in Côte d'Ivoire, West Africa. J Ethnobiol Ethnomed 15(5):1–12

Srivastava B, Dwivedi AK, Pandey VN (2011) Ethnobotanical survey, distribution and utilization of *Termitomyces* species in Gorakhpur forest division. PSF 1(3):28–33

Stoyneva-Gärtner MP, Uzunov BA, Dimitrova P (2017) Jelly-like algae and fungi used as food in Bulgaria. Acta Sci Nutr Health 1(1):51–54

Stryamets N, Elbakidze M, Ceuterick M, Angelstam P, Axelsson R (2015) From economic survival to recreation: contemporary uses of wild food and medicine in rural Sweden, Ukraine and NW Russia. J Ethnobiol Ethnomed 11:53):1–53)19

Sung JM, Lee HK, Yoo YJ, Choi YS, Kim SH, Kim YO, Sung GH (1998) Classification of *Cordyceps* species based on protein banding pattern. Kor J Mycol 26:1–7

Tantengco OAG, Ragragio EM (2018) Ethnomycological survey of macrofungi utilized by Ayta communities in Bataan, Philippines. Curr Res Environ Appl Mycol 8(1):104–108

Tanti B, Lisha G, Sharma GC (2011) Wild edible fungal resources used by ethnic tribes of Nagaland, India. Indian J Tradit Know 10:512–515

Teferi Y, Muleta D, Woyessa D (2013) Mushroom consumption habits of Wacha Kebele resident, Southwestern Ethiopia. Global Res J Agric Biol Sci 4(1):6–66

Teke NA, Kinge TR, Bechem E, Nji TM, Ndam LM, Mih AM (2018) Ethnomycological study in the Kilum-Ijim mountain forest, Northwest region, Cameroon. J Ethnobiol Ethnomed 14(1):25

Thangaraj R, Raj S, Renganathan K (2017) Wound healing effect of King Alfred's mushroom (*D.concentrica*) used by tribes of Sirumalai hills, Tamil Nadu, India. Int J Pharm Pharm Sci 9(7):161–164

Trutmann P, Holgado ME, Quispe A, Luque A (2012) Native mushrooms, local knowledge, and potential for food and health in the Peruvian Andes: update 2012. Annual report 2012, Global Mountain Action, pp 1–6

Tuno N (2001) Mushroom utilization by the Majangir, an Ethiopian tribe. Mycologist 15(2):78–79

Valverde ME, Hernández-Pérez T, Paredes-López O (2015) Edible mushrooms: improving human health and promoting quality life. Int J Microbiol 2015:1–14

Villarreal L, Perez-Moreno J (1989) Los hongos comestibles silvestres de Mexico, un enfoque integral. Micologia Neotropica Aplicada 2:77–114

Vishwakarma P, Tripathi NN (2019) Ethnomacrofungal study of some wild macrofungi used by local people of Gorakhpur district, Uttar Pradesh. Indian J Nat Prod Resour 10(1):81–89

Wasson RG (1971) Soma: divine mushroom of immortality, Ethnomycology studies No. 1, Harvest special. Harcourt Brace Jovanovich, New York

Zent EL, Zent S, Iturriaga T (2004) Knowledge and use of fungi by a mycophilic society of the Venezuelan Amazon. Econ Bot 58(2):214–226

Chapter 5
Toxigenic Fungi

Majority of the macrofungi are edible/medicinal, while some species are toxigenic causing fatal accidents annually (Wu et al. 2019). Of all the identified fungi in the world, less than 1% are toxigenic (Chang 2008). The poisoning of macrofungi in humans is known since time immemorial and is mentioned in ancient writings like "Rigveda" (at least 3500 B.C.) and "Atharvaveda" (at least 1500 B.C.) (Verma et al. 2014). Avoidance, insufficient knowledge, and misidentification lead to the consumption of toxic macrofungi causing various health hazards, such as vomiting, nausea, stomachache, gastroenteritis, diarrhea, hepatotoxicity, nephrotoxicity, and neurotoxicity, may cause rhabdomyolysis symptoms and erythromelalgia syndrome, and even sometimes lead to mortality (Erguven et al. 2007; Vişneci et al. 2019). Therefore, grasping, preservation, and transmission of ethnomycological knowledge are critical to prevent fungal poisoning or mycetism (Kim and Song 2014). Most of the time, accidental toxicity happens because of misidentification leading to ingestion of toxigenic fungi. The lack of data on toxigenic fungi and information pertaining to toxicological profiles of these fungi contributes more towards their consumption (White et al. 2003; Flesch and Saviuc 2004). Therefore, correct identification of the collected specimens before utilization is prerequisite to avoid toxicity. Some of the common differences usually observed between edible and toxigenic species are given in Table 5.1. Moreover, the national governments in different countries release guidelines and enact legislations for safe commerce and utilization of macrofungi (Peintner et al. 2013).

Most of the toxigenic species occur in the genera *Amanita* (*Amanitaceae*), *Clitocybe* (*Physalacriaceae*), *Cortinarius* (*Cortinariaceae*), *Conocybe* (*Bolbitiaceae*), *Entoloma* (*Entolomataceae*), *Gyromitra* (*Discinaceae*), *Hebeloma* (*Strophariaceae*), *Inocybe* (*Strophariaceae*), *Lepiota* (*Agaricaceae*), *Omphalotus* (*Omphalotaceae*), *Psilocybe* (Strophariaceae), and *Russula* (*Russulaceae*), etc. as is evident from the data provided in Table 5.2 (Fig 5.1).

© The Editor(s) (if applicable) and The Author(s), under exclusive license to Springer Nature Switzerland AG 2020
U. Azeem et al., *Fungi for Human Health*,
https://doi.org/10.1007/978-3-030-58756-7_5

Table 5.1 Differences between edible and toxigenic fungi (Ramírez-Terrazo et al. 2014)

Edible fungi	Toxigenic fungi
Fruit in familiar substratum or dead and decaying tree trunk. Have a clear distinct ring on the stalk at maturity	Grow generally on partially decomposing animal dung (coprophilous fungi) without ring and black gills. Net mushrooms (the macrofungi having net on the sporocarps) are considered highly poisonous
Have familiar and pleasant odor	With peculiar and unpleasant odor
Sporocarps or gills become red brown when harvested	Colored sporocarps with black spots on gills and turn black on breakage
Stipe becomes brown on breakage and releases brown color in the water	Stipe turns black on breakage or when picked up from the soil. Sporocarps on breakage release mucilage substance from stalk and turn black or blue in a mixture of salt water with lemon juice

Many myths are associated with the toxicity of macrofungi. It is believed that some macrofungal species are tested by snakes making them poisonous. In Western Burundi (Africa), it is famous that if once poisonous mushroom is collected, it should be kept back to its place to avoid any mishappening (Atri and Mridu 2018). The scientific cause of toxicity is the presence of different classes of toxins present in macrofungal species. The diseases caused by intake of mycotoxins are called mycotoxicoses. Toxins such as coprine (*Coprinus atramentarius*), acromelic acid (*C. acromelalga*), agaritine (*A. bisporus*), amatoxin (different species of *Amanita*), gyromitrin (*G. esculenta*), ibotenic acid and muscimol (*A. muscaria* and *A. pantherina*), muscarine (*Clitocybe serussata, C. dealbata, C. phyllophila, C. rivulosa*, and *A. muscaria*), orellanine (*P. ostreatus* and *C. orellanus*), ostreolysin (*P. ostreatus*), phallotoxin (*A. phalloides*), psilocybin and psilocin (*Psilocybe cubensis*), etc. cause toxicity following diverse routes inside the human body. Some species in the genera *Inocybe* and *Clitocybe* are more toxic as compared to their counterparts in *Entoloma* and *Mycena* because the concentration of toxins is higher in the former genera. Muscarine (a toxin) is 0.1 and 0.33% in *Inocybe* and *Clitocybe* spp., respectively, while it is 0.0003% in *A. muscaria* accounting for less toxicity of the latter species than the species in the former genera (Gupta 2018). Toxins present in some species are also beneficial (Nieminen et al. 2006). For instance, *Agaricus xanthodermus* is a poisonous mushroom but possesses medicinal properties such as anti-atherosclerosis, antioxidant, and antimicrobial effects (Özaltun and Sevindik 2020). Accidents of mild poisoning also occur with the ingestion of edible/medicinal species occasionally. This possibility of toxicity can be avoided through some recipes such as a drop of lemon is added to the basidiocarps of *Termitomyces* while cooking (Semwal et al. 2014). In case of *G. floccosus*, scales are removed from the cap and

Table 5.2 The most commonly occurring species of toxigenic fungi in the world

Current scientific name	Family	Reference
Agaricus moelleri, A. bresadolanus, A. hondensis, A. silvaticus, A. squarrosus, A. xanthodermus, Chlorophyllum molybdites, and *Lepiota cristata*	*Agaricaceae*	Joval et al. (1996), Demürel and Uzun (2004), Beug (2010), Lima et al. (2012), Edwards and Leech (2014), Liu et al. (2018), Rothman (2018), Kotowski et al. (2019) and Vishwakarma et al. (2019)
Amanita arocheae, A. bisporigera, A. citrine, A. echinocephala, A. eliae, A. flavoconia, A. gemmata, A. muscaria, A. pantherina, A. phalloides, A. rubescens, A. solitaria, A. smithiana, A. verna, and *A. virosa*	*Amanitaceae*	Montoya et al. (2003), Demürel and Uzun (2004), Borovicka (2006), Yardan et al. (2008), Murati and Rexhepi (2018), Ruan-Soto (2018), Kotowski et al. (2019) and White et al. (2019)
Bolbitius sp. and *Conocybe albipes*	*Bolbitiaceae*	Hallen et al. (2003) and Haro-Luna et al. (2019)
B. edulis, B. miniato-pallescens, Leccinum aurantiacum, Rubroboletus satanas, Suillellus luridus, and *Xerocomus truncatus*	*Boletaceae*	Montoya et al. (2003), Demürel and Uzun (2004), Ruan-Soto (2018), White et al. (2019), Haro-Luna et al. (2019) and Kotowski et al. (2019)
Cortinarius orellanus, C. speciosissimus, and *C.* sp.	*Cortinariaceae*	Lima et al. (2012) and White et al. (2019)
G. esculenta	*Discinaceae*	Yardan et al. (2008)
Entoloma incanum, E. sericellum, E. sericeoides, and *E. sinuatum*	*Entolomataceae*	Demürel and Uzun (2004)
Ganoderma neojaponicum	*Ganodermataceae*	White et al. (2019)
Corallium formosum	*Gomphaceae*	Ruan-Soto (2018)
Hygrophoropsis aurantiaca	*Hygrophoropsidaceae*	Desjardin et al. (2015) and Kotowski et al. (2019)
Podostroma cornu-damae and *P. rubicundulus*	*Hypocreaceae*	Demürel and Uzun (2004) and White et al. (2019)
Inocybe asterospora, I. dulcamara, I. erubescens, I. fastigiata f. *subcandida, I. lilacina, I. napipes, I. pallidicremea, I. patouilladii, I. radiate, I. repanda,* and *I. rimosa*	*Inocybaceae*	Demürel and Uzun (2004), Lima et al. (2012) and Ruan-Soto (2018)
Pleurocybella porrigens	*Marasmiaceae*	Lima et al. (2012) and White et al. (2019)
Morchella sp.	*Morchellaceae*	White et al. (2019)
Mycena pura	*Mycenaceae*	Verma et al. (2014)
Omphalotus japonicus, O. olearius, and *O. olivascens*	*Omphalotaceae*	Verma et al. (2014) and Bal et al. (2016), Sugano et al. (2017)
Paxillus involutus	*Paxillaceae*	White et al. (2019)

(continued)

Table 5.2 (continued)

Current scientific name	Family	Reference
Sarcosphaera crassa	*Pezizaceae*	Demürel and Uzun (2004), Dahlberg and Croneborg (2006) and Tedersoo et al. (2006)
Hapalopilus nidulans	Phanerochaetaceae	Lima et al. (2012) and White et al. (2019)
Pluteus salicinus	Pluteaceae	Demürel and Uzun (2004) and Justo et al. (2014)
Panaeolus antillarum, P. foenisecii, and *P. olivaceus*	*Psathyrellaceae*	Demürel and Uzun (2004), Desjardin and Perry (2017) and Haro-Luna et al. (2019)
Ramaria abietina and *R. apiculata*	*Ramariaceae*	Montoya et al. (2003)
Lactarius aurantiacus, L. luculentus, L. piperatus, L. torminosus, Russula betularum, R. emetica, R. fellea, R. fragilis, R. luteotacta, R. subnigricans	*Russulaceae*	Montoya et al. (2003), Demürel and Uzun (2004), Phillips (2006), Ruan-Soto (2018), White et al. (2019), Kotowski et al. (2019) and Verma et al. (2019)
Scleroderma areolatum	*Sclerodermataceae*	Ruan-Soto (2018)
Galerina marginata, Hebeloma fusipes, H. mesophaeum, H. sinapizans, Hypholoma fasciculare, Psilocybe baeocystis, P. bohemica, P. cubensis, P. mexicana, and *P. semilanceata*	*Strophariaceae*	Lima et al. (2012)
Tapinella atrotomentosa	*Tapinellaceae*	Kotowski et al. (2019)
Clitocybe amoenolens, C. acromelalga, C. dealbata, C. phyllophila, C. rivulosa, Leucocybe candicans, Tricholoma equestre, T. scalpturatum, and *T. ustale*	*Tricholomataceae*	Demürel and Uzun (2004), Lima et al. (2012), Nakajima et al. (2013) and White et al. (2019)

Fig. 5.1. Some common toxigenic macrofungal species known for causing mycetism (1) *Agaricus xanthodermus*, (2) *A. muscaria*, (3) *A. pantherina*, (4) *A. rubescens*, (5) *Boletus edulis*, (6) *Chalciporus piperatus*, (7) *Coprinopsis atramentaria*, (8) *Entoloma sinuatum*, (9) *Hebeloma sinapizans*, (10) *Hygrophoropsis aurantiaca*, (11) *Inocybe erubescens*, (12) *Lactarius torminosus*, (13) *Panaeolus foenisecii*, (14) *P. olivaceus*, (15) *Pluteus salicinus*, (16) *Psilocybe semilanceata*, (17) *Ramaria abietina*, (18) *Rubroboletus satanas*, (19) *Russula emetica*, (20) *Sarcosphaera crassa*, (21) *S. luridus*, (22) *Tricholoma ustale*. (https://www.first-nature.com; https://www.gbif.org; https://en.wikipedia.org)

Fig. 5.1 (continued)

the veins of the sporocarps before cooking to get rid of bitterness (Lincof and Mitchel 1977). Thus, proper identification and edibility studies are necessary for correct scientific valorization of macrofungal taxa as food and medicine.

References

Atri NS, Mridu C (2018) Mushrooms-some ethnomycological and sociobiological aspects. Kavaka 51:11–19

Bal A, Anil M, Yilmaz I, Akata I, Atilla OD (2016) An outbreak of non-fatal mushroom poisoning with *Omphalotus olearius* among Syrian refugees in Izmir, Turkey. Toxin Rev 35:1–2):1–3

Beug MW (2010) An overview of mushroom poisonings in North America. Mycophile 45(2):4–5

Borovicka J (2006) Notes on *Amanita strobiliformis* and related species of the section Lepidella. Mykol Sborn 83(2):43–46

Chang ST (2008) Overview of mushroom cultivation and utilization asnfunctional foods. In: Cheung PCK (ed) Mushrooms as functional foods. Hoboken, Wiley, pp 1–33

Dahlberg A, Croneborg H (2006) The 33 threatened fungi in Europe (Nature and Environment), No 136. Council of Europe, Strasbourg, pp 104–108

Demürel K, Uzun Y (2004) Some poisonous fungi of East Anatolia. Turk J Bot 28(1–2):215–219

Desjardin DE, Perry BA (2017) *Panaeolus antillarum* (Basidiomycota, Psathyrellaceae) from wild elephant dung in Thailand. Curr Res Environ Appl Mycol 7(4):275–281

Desjardin DE, Wood MG, Stevens FA (2015) California mushrooms: the comprehensive identification guide. Timber Press, Portland

Edwards A, Leech T (2014) *Agaricus Bresadolanus* – a toxic mushroom. Field Mycol 15(4):113–114

Erguven M, Yilmaz O, Deveci M, Aksu N, Dursun F, Pelit M, Cebeci N (2007) Mushroom poisoning. Indian J Pediatr 74:847–852

Flesch F, Saviuc P (2004) Intoxications par les champignons: principaux syndromes et traitement. EMC Med 1(1):70–79

Gupta RC (2018) Veterinary toxicology: basic and clinical principles, 3rd edn. Academic Press, p 1238

Hallen HE, Watling R, Adams GC (2003) Taxonomy and toxicity of *Conocybe lactea* and related species. Mycol Res 107(8):969–979

Haro-Luna MX, Ruan-Soto F, Guzmán-Dávalos L (2019) Traditional knowledge, uses, and perceptions of mushrooms among the Wixaritari and mestizos of Villa Guerrero, Jalisco, Mexico. IMA Fungus 10(16):1–14

Joval E, Kroeger P, Towers N (1996) Hydroquinone: the toxic compound of *Agaricus hondensis*. Planta Med 62(2):185–185

Justo A, Malysheva E, Bulyonkova T, Vellinga EC, Cobian G, Nguyen NH, Minnis AM, Hibbett DS (2014) Molecular phylogeny and phylogeography of Holarctic species of *Pluteus* section *Pluteus* (Agaricales: Pluteaceae), with description of twelve new species. Phytotaxa 180(1):1–85

Kim H, Song MJ (2014) Analysis of traditional knowledge for wild edible mushrooms consumed by residents living in Jirisan National Park (Korea). J Ethnopharmacol 153(1):90–97

Kotowski MA, Pietras M, Łuczaj L (2019) Extreme levels of mycophilia documented in Mazovia, a region of Poland. J Ethnobiol Ethnomed 15(12):1–19

Lima ADL, Fortes RC, Garbi Novaes MRC, Percário S (2012) Poisonous mushrooms: a review of the most common intoxications. Nutr Hosp 27(2):402–408

Lincof G, Mitchel DH (1977) Toxic and hallucinogenic mushroom poisoning. A handbook for physicians and mushroom hunters. Litton Educational Publishing Inc., New York, US

Liu D, Cheng H, Bussmann RW, Guo Z, Liu B, Long C (2018) An ethnobotanical survey of edible fungi in Chuxiong City, Yunnan, China. J Ethnobiol Ethnomed 14(42):1–10

Montoya A, Hernandez-Totomoch O, Estrada-Torres A, Kong A, Caballero J (2003) Traditional knowledge about mushrooms in a Nahua community in the state of Tlaxcala, Mexico. Mycologia 95(5):793–806

Murati E, Rexhepi B (2018) Edible and poisonous mushrooms. ECOTEC-J Sci Env Technol 1(1):41–44

Nakajima N, Ueda M, Higashi N, Katayama Y (2013) Erythromelalgia associated with *Clitocybe acromelalga* intoxication. Clin Toxicol 51(5):451–454

Nieminen P, Kirsi M, Mustonen AM (2006) Suspected myotoxicity of edible wild mushrooms. Exp Biol Med 231(2):221–228

Özaltun B, Sevindik M (2020) Evaluation of the effects on atherosclerosis and antioxidant and antimicrobial activities of *Agaricus xanthodermus* poisonous mushroom. Eur Res J 6(2):1–7

Peintner U, Schwarz S, Mesic A, Moreau PA, Moreno G, Saviuc P (2013) Mycophilic or mycophobic? legislation and guidelines on wild mushroom commerce reveal different consumption behaviour in European countries. PLoS ONE 8(5):1–10

Phillips R (2006) Mushrooms: A comprehensive guide to mushroom identification. Pan Macmillan Ltd, London

Ramírez-Terrazo A, Montoya A, Caballero J (2014) Una mirada al conocimiento tradicional sobre los hongos tóxicos en México. In: La etnomicología en México, estado del arte. CONACYT, UAEH, and UNAM, Mexico, pp 113–141

Rothman S (2018) Edible macrofungi of Namibia's thorn bush savanna bioregion and their potential for sustainable development. Master's thesis. Natural Resource Management, Raseborg

Ruan-Soto F (2018) Sociodemographic differences in the cultural significance of edible and toxic mushrooms among Tsotsil towns in the highlands of Chiapas. Mexico. J Ethnobiol Ethnomed 14(32):1–19

Semwal KC, Stephenson SL, Bhatt VK, Bhatt RP (2014) Edible mushrooms of the Northwestern Himalaya, India: a study of indigenous knowledge. distribution and diversity. Mycosphere 5(3):440–461

Sugano Y, Sakata K, Nakamura K, Noguchi A, Fukuda N, Suzuki T, Kondo K (2017) Rapid identification method of *Omphalotus japonicus* by Polymerase Chain Reaction-Restriction Fragment Length Polymorphism (PCR-RFLP). Shokuhin eiseigaku zasshi. J Food Hyg Soc Jpn 58(3):113–123

Tedersoo L, Hansen K, Perry BA, Kjøller R (2006) Molecular and morphological diversity of Pezizalean ectomycorrhiza. New Phytol 170(3):581–596

Verma N, Bhalla A, Kumar S, Dhimany RK, Chawlay YK (2014) Wild mushroom poisoning in North India: Case series with review of literature. J Clin Exp Hepatol 4(4):361–365

Verma RK, Pandro V, Rao GR (2019) Three records of *Russula* mushroom from Sal forest of Central India. Int J Curr Microbiol Appl Sci 8(2):445–455

Vishwakarma P, Singh P, Tripathi NN (2019) Biodiversity of macrofungi from Gorakhpur district (UP), India. NeBIO 10(1):5–11

Vişneci EF, Acar D, Özdamar EN, Güven M, Patat M (2019) Mushroom poisoning cases from an emergency department in Central Anatolia: Comparison and evaluation of wild and cultivated mushroom poisoning. Eurasian J Emerg Med 18(1):28–33

White J, Warrell D, Eddleston M, Currie BJ, Whyte IM, Isbister GK (2003) Clinical toxinology: where are we now? J Toxicol Clin Toxicol 41(3):263–726

White J, Weinsteina SA, Harob LD, Bédryc R, Schaperd A, Rumacke BH, Zilkerf T (2019) Mushroom poisoning: A proposed new clinical classification. Toxicon 157:53–65

Wu F, Zhou LW, Yang ZL, Bau T, Li TH, Dai YC (2019) Resource diversity of Chinese macrofungi: Edible, medicinal and poisonous species. Fungal Divers 98(1):1–76

Yardan T, Eden AO, Baydin A, Arslan B, Vural K (2008) Mushroom poisonings. Ondokuz Mayis Uni Med J 25(2):75–83

Websites Followed

https://en.wikipedia.org
https://www.first-nature.com
https://www.gbif.org

Chapter 6
Nutritional Significance

An array of nutrients like carbohydrates, proteins, amino acids, fats, fatty acids, fibers, minerals, and vitamins are present in macrofungi. These nutrients make them an ideal candidate to be used in nutraceutical formulations. Macrofungi are low in calories and cholesterol and are rich in antioxidants adding more to their health-promoting qualities (Akata et al. 2013; Chauhan et al. 2014; Sinanoglou et al. 2014; Ghosh 2016; Reis et al. 2017; Üstün et al. 2018; Tietel and Masaphy 2018). The amount of proteins, carbohydrates, fats, fibers, moisture, and ashes lies approximately in the range of 31.40–49.05%, 41.25–65%, 0.25–1.08%, 6.41–17.97%, 54.75–76.97%, and 2.64–10.34%, respectively (Vishwakarma et al. 2016). This range of mycochemicals varies with the species under investigation. Polysaccharides in mushrooms are made up of many monosaccharides such as glucose, galactose, fructose, xylose, mannose, fucose, rhamnose, arabinose, trehalose, and mannitol (Valverde et al. 2015). Fungi are known for their high protein content consisting of various types of amino acids. Amino acid profiling of *A. hemibapha*, *Boletinus pinetorus*, *Baorangia bicolor*, *B. speciosus*, *B. sinicus*, *B. craspedius*, *B. griseus*, *B. ornatipes*, *Suillus placidus*, *T. microcarpus*, *Tricholoma terreum*, *Tricholomopsis lividipileata*, and *Xerocomus* has been performed through pre-column derivatization reversed-phase high-performance liquid chromatography (RP-HPLC). These species have been found containing all the 20 amino acids, aspartic acid, glutamic acid, serine, glycine, alanine, praline, cysteine, valine, methionine, phenylalanine, isoleucine, leucine, lysine, histidine, threonine, asparagine, glutamine, arginine, tyrosine, and tryptophan, but each species has presented its own amino acid profile. The total free amino acid content varies from 1462.6 to 13,106.2 mg/100 g. Alanine, cysteine, glutamine, and glutamic acid represent the most abundant amino acids present in these species (Sun et al. 2017). Fats occur in low amounts in macrofungi and are rich in both saturated and unsaturated fatty acids such as cis-11-eicosenoic acids, cis-10-pentadecenoic acid, linoelaidic acid, linoleic acid, myristic acid, oleic acid, palmitic acid, pentadecanoic acid, etc. as analyzed by gas chromatography coupled with flame ionization detector (GC-FID) and chromatography coupled with mass

U. Azeem et al., *Fungi for Human Health*, https://doi.org/10.1007/978-3-030-58756-7_6

spectroscopy (GC-MS). Unsaturated fatty acids dominate fats in macrofungi and mainly consist of linoleic acid (10–51%) (Türkekul et al. 2017; Bengu 2020). Besides, carbohydrates, proteins, fats, various minerals, and vitamins have also been found in many macrofungal taxa (Thatoi and Singdevsachan 2014; Nakalembe et al. 2015). Mineral profiling performed through wavelength dispersive X-ray fluorescence (WDXRF) technique has detected the presence of several mineral nutrients in the dry ash made from basidiocarps of *Phellinus* mushrooms, namely, *Ph. conchatus*, *Ph. rimosus*, *Ph. igniarius*, *Ph. gilvus*, and *Ph. nigrolimitatus* (Chenghom et al. 2010). Several mineral nutrients such as Cu, Zn, Fe, Mn, Co, Cd, Ni, and Pb have been determined in 24 macrofungal species (Yilmaz et al. 2003). Vitamins such as A, B, C, D, and E have been detected in several fungal species (Wani et al. 2010; Phillips et al. 2011; Phillips and Rasor 2013; Gaitán-Hernández et al. 2019). The liquid chromatography-mass spectroscopy (LC-MS) reported the presence of several phenolics and flavonoids in the mature and immature summer truffle (*T. aestivum*) (Shah et al. 2020). Since macrofungi contain a variety of nutrients such as carbohydrates, proteins, minerals, vitamins, amino acids, fatty acids, etc. (Table 6.1), the consumption of these species has enhanced either as whole mushrooms or extracted supplements. The DSs of mushrooms boost the innate immunity and help the body to fight against various diseases. Edible mushroom supplements enriched with nutrients have also been screened for their pharmaceutical potential especially against cancer. *A. bisporus* (white button mushroom) nutritional supplements have been found to strengthen the innate immunity which proves helpful to fight against tumors and viruses by promoting natural killer (NK) cell potential probably mediated through enhanced production of interferon gamma (IFNγ) and tumor necrosis factor alpha (TNFα) (Wu et al. 2007). A dietary supplement, MycoPhyto® Complex (MC), is a mixture of mycelia from *A. blazei*, *G. frondosa*, *G. lucidum*, *T. versicolor*, *O. sinensis*, *P. umbellatus*, and β-1,3-glucan extracted from *Saccharomyces cerevisiae*. This mixture possesses antiproliferative properties and arrests the cell cycle at G2/M phase of human breast cancer cells MDA-MB-231 through the suppression of cell cycle regulatory genes (ANAPC2, ANAPC2, BIRC5, Cyclin B1, Cyclin H, CDC20, CDK2, CKS1B, Cullin 1, E2F1, KPNA2, PKMYT1, and TFDP1). The anti-invasive effect of MC is due to the inhibition of production of urokinase plasminogen activator (uPA) from MDA-MB-231 cells (Jiang and Sliva 2010). It has been reported that dietary supplementation of *P. eryngii* powder (1.5 and 2%) improves the humoral innate immune responses, antibacterial efficiency of skin mucus, and growth performance of Koi fish (Safari and Sarkheil 2018). The cultivated basidiocarps of *G. lucidum* have been found to accumulate elements from the substrate and can be utilized for the production of supplements rich in mineral nutrients (Rzymski et al. 2016). *Taiwanofungus camphoratus* mushroom is parasite on *Cinnamomum* kanehirae, and the cold water-soluble polysaccharide (galactomannan-repeated, molecular weight >70 kDa) isolated from this mushroom exhibits immuno-enhancing properties. The DSs of this polysaccharide show immunomodulation in mouse macrophages and human dendritic cells. It can be used as adjuvant in immunotherapy and vaccination (Perera et al. 2018). Mushroom polysaccharides have also been developed and exploited as functional foods such as the

Table 6.1 Nutritional composition of some health-benefiting macrofungi

Current name	Nutraceutical components	Reference
Auricularia nigricans, Bovista pusilla, Calvatia gigantea, L. squarrosulus, Macrocybe lobayensis, P. ostreatus, L. tuber-regium, Psathyrella atroumbonata, S. commune, T. microcarpus, T. globulus, and *Volvariella esculenta*	Carbohydrates, sugars (rhamnose, fructose, mannitol, sucrose, and trehalose), glycogen, fibers, ashes, fats, proteins, moisture, and minerals (Ca, Cu, Fe, K, Mg, Mn, Na, P, and Zn)	Gbolagade et al. (2006)
A. bisporus, A. silvaticus, A. sylvicola, B. edulis, C. gambosa, C. cibarius, Craterellus cornucopioides, and *Marasmius oreades*	Carbohydrates, sugars, proteins, fibers, ashes, fats, fatty acids, phenolics, flavonoids, carotenoids, and ascorbic acid	Barros et al. (2008)
A. mellea, A. tabesceus, Boletus aureus, C. cibarius, Fistulina hepatica, Hygrophorus russula, Lepista nuda, Ramaria largentii, and *Russula delica* var. *chloroides*	Carbohydrates, proteins, fats, moisture, ashes, and fibers	Ouzouni et al. (2009)
Chlorophyllum molybditis, Lentinus subnudus, Marasmus sp., and *L. tuberregium*	Carbohydrates, proteins, fat, moisture, fibers, and ashes	Adedayo et al. (2010)
C. cibarius, Clitocybe odora, H. erinaceus, Laccaria laccata, L. deliciosus, Lepista nuda, L. saeva, M. procera, and *P. ostreatus*	Carbohydrates, proteins, fibers, ashes, fats, minerals (Ca, Cu, Fe, K, Mg, Mn, and Na), alkaloids, flavonoids, saponins, and tannins	Egwim et al. (2011)
A. auricula-judae, G. frondosa, L. squarrosulus, Lentinus sajor-caju, L. tuber-regium, P. ostreatus, Pleurotus roseolus, T. microcarpus, Termitomyces heimii, and *V. volvacea*	Carbohydrates, proteins, fats, moisture, ashes, and minerals	Johnsy et al. (2011)
A. bisporus, L. edodes, P. ostreatus, P. eryngii, and *F. velutipes*	Carbohydrates, proteins, moisture, fatty acids, sugars, and tocopherol	Reis et al. (2012)
Ramaria aurea	Carbohydrates, proteins, fats, amino acids, crude fiber, and minerals	Manjula and Krishnendu (2012)
Pleurotus citrinopileatus	Fibers, proteins, fats, moisture, minerals, amino acids, vitamins (B2, B3, B5, B12, and vitamin A)	Musieba et al. (2013)
L. squarrosulus, Russula albonigra, Tricholoma giganteum	Proteins, lipids, moisture, and fibers	Giri et al. (2013)
P. cornucopiae, P. ostreatus, P. pulmonarius, and *P. sapidus*	Carbohydrates, proteins, fats, ashes, moisture, phenolics, flavonoids, alkaloids, saponins, steroids, phlobatannins, and anthraquinones	Adebayo et al. (2014)

(continued)

Table 6.1 (continued)

Current name	Nutraceutical components	Reference
A. bisporus, A. caesarea, B. edulis, C. cibarius, F. velutipes, F. hepatica, L. deliciosus, L. edodes, M. procera, M. esculenta, P. eryngii, and *P. ostreatus*	Ergosterol	Barreira et al. (2014)
Agaricus brasiliensis, C. comatus, Cordyceps militaris, F. velutipes, G. lucidum, G. frondosa, H. erinaceus, L. edodes, O. sinensis P. ostreatus, T. versicolor, and *T. fuciformis*	Carbohydrates, proteins, fats, fatty acids, amino acids, moisture, minerals (Ag, Al, As, B, Ba, Be, Ca, Cd, Co, Cr, Cu, Fe, Hg, K, Li, Mg, Mn, Mo, Na, Ni, P, Pb, S, Se, Sr, Ti, V, W, and Zn), and ashes	Cohen et al. (2014)
Polyporus tenuiculus, T. clypeatus, T. microcarpus, T. tyleranus, and *Volvariella gloiocephala*	Carbohydrates, proteins, moisture, minerals (K, P, Fe Ca, Cu, Mg, Mn, Na, Se, Zn), vitamins (thiamin, vitamin C, folic acid, and niacin), and ashes	Nakalembe et al. (2015)
Termitomyces robustus and *L. squarrosulus*	Carbohydrates, proteins, fats, mineral nutrients (Ca, Cu, Fe, Hg, K, Mg, Mn, Na, P, Pb, and Zn), moisture, amino acids, vitamins (vitamin A, vitamin C, thiamine, and tocopherol), phenolic compounds (gallic acid, catechin, chlorogenic acid, caffeic acid, ellagic acid, epicatechin, rutin, isoquercitrin, quercitrin, quercetin, and kaempferol), tannins saponins, terpenoids, and ashes	Borokini et al. (2016)
Aleurodiscus vitellinus, Cyttaria hariotii, Cortinarius magellanicus, Fistulina hepatica, Grifola gargal, Hydropus dusenii, L. nuda, and *Ramaria patagonica*	Carbohydrates, proteins, sugars, fats, fatty acids, tocopherols, organic acids, phenolic compounds and ashes	Toledo et al. (2016)
Fuscoporia torulosa, Phellinus allardii, P. gilvus, P. fastuosus, and *P. sanfordii*	Carbohydrates, proteins, phenols, flavonoids, tannins, alkaloids, sugars, glycosides, terpenoids, steroids, ashes, and moisture	Azeem (2017)
H. repandum, L. deliciosus, and *Tricholoma equestre*	Carbohydrates, sugars (glucose, mannitol, trehalose), proteins, fats, mineral nutrients (Na, K, Mg, Ca, Cu, Fe, Zn), amino acids, ashes, and moisture	Jedidi et al. (2017)
Twelve wild strains of *Ganoderma* sp.	Carbohydrates (free sugars: rhamnose, fructose, mannitol, sucrose, trehalose, and β-glucans) proteins, fats, fatty acids, organic acids, phenolic compounds, minerals, vitamins, and ashes	Obodai et al. (2017)
G. lucidum L. edodes, P. eryngii and *P. ostreatus*, and *V. volvacea*	Carbohydrates, proteins, fats, moisture, fibers, ashes, and minerals (Ca, Cu, Fe, Mg, Mn and P Zn), as well as micronutrient	Salamat et al. (2017)

(continued)

Table 6.1 (continued)

Current name	Nutraceutical components	Reference
Clavaria amoena, C. coralloides, C. fragilis, C. rosea, C. vermicularis, Clavulina amethystina Ramaria aurea, R. botrytis, R. flava, R. flavescens, R. rubripermanens, and *R. stricta*	Carbohydrates, proteins, ashes, moisture, minerals (Ca, Cu, Fe, K, Mg, and Na), fatty acids, amino acids, phenolics, tocopherols, anthocyanins, and carotenoids	Sharma and Gautam (2017)
Agaricus campestris, Amanita vaginata, Astraeus hygromatricus, Leucopaxillus sp., *P. ostreatus, Russula delica, S. commune,* and *T. heimii*	Carbohydrates, proteins, fats, moisture, fibers, and ashes	Singha et al. (2017)
A. campestris, Helvella imbricatum, P. ostreatus and *S. crispa*	Proteins, fats, moisture, and fibers	Ullah et al. (2017)
P. ostreatus, L. edodes, and *M. procera*	Carbohydrates, proteins, fats, moisture, ashes, α-glucans, β-glucans, and minerals (Ag, Al, As, B, Ba, Bi, Ca, Cd, Co, Cr, Cu, Fe, Ga, Ge, In, K, Li, Mg, Mn, Mo, Na, Ni, P, Pb, Pt, Sb, Se, Si, Sr, Ti, and Zn)	Kolundžić et al. (2018)
Amanita rubescens var. rubescens, B. edulis, G. arenicola, Morchella deliciosa, and *S. crispa*	Proteins, phenols, fibers, ashes, and minerals (Cu, Fe, Mn, and Zn)	Lalotra et al. (2018)
Picoa juniperi, Terfezia boudieri, T. claveryi, T. sp., *Tirmania nivea,* and *T. pinoyi*	Carbohydrates, proteins, fats, moisture, fibers, and ashes	Owaid (2018)
A. auricula-judae, G. applanatum, G. lucidum, P. ostreatus, and *S. commune*	Carbohydrates (glucose, mannose, xylose, and galactose), proteins, amino acids, fats, fibers, ashes, and minerals (Ca, Cr, Cu, Fe, Mg, P, and Zn)	Pathania and Chander (2018)
B. edulis, C. cornucopioides, G. lucidum, H. erinaceus, L. deliciosus, L. sulphureus, L. edodes, M. oreades, M. conica, P. ostreatus, R. botrytis, and *T. terreum*	Carbohydrate, proteins, amino acids, phenols, flavonoids, sugars (glucose, fructose, and sucrose), and elements (Ag, Al, As, Ba, Bi, Br, Ca, Cd, Cl, Co, Cr, Cu, Fe, Ga, Ge, Hg, Hf, I, K, La, Mg, Mn, Mo, Na, Nb, Nd, Ni, P, Pb, Pr, Rb, S, Sb, Se, Si, Sn, Sr, and Ta)	Turfan et al. (2018)
A. auricula-judae, A. nigricans, Lactifluus piperatus, L. sulphureus, L. edodes, L. squarrosulus, L. sajor-caju, L. tigrinus, S. commune, and *T. heimii*	Carbohydrates, proteins, fibers, moisture, ashes, sugars, phenols, and flavonoids	Ao and Deb (2019)
A. auricula-judae, A. fuscosuccinea, A. nigricans, and *A. thailandica*	Carbohydrates, proteins, fats, fibers, ashes, sugars, amino acids, and minerals (Ca, Cr, Cu, Fe, K, Mg, Mn, Na, Ni. P, and Zn)	Bandara et al. (2019)
T. aestivum	Phenolics, flavonoids, and fatty acids	Shah et al. (2020)

(continued)

Table 6.1 (continued)

Current name	Nutraceutical components	Reference
Infundibulicybe geotropa	Phenols (catechin, chlorogenic acid, and coumaric acid) and minerals (Fe, Cu, Ni, Pb, Zn)	Sevindik et al. (2020)

lentinan in *L. edodes*, schizophyllan in *S. commune*, pleuran in *Pleurotus* species, calocyban in *Calocybe indica*, and ganoderan in *G. lucidum* (Villares et al. 2012; Badalyan 2016). They function as dietary fibers and interact with microbiota residing in the gastrointestinal tract bringing changes in the gut microbiota abundance which in turn impacts the host health (Jayachandran et al. 2017). Basically, gut microbiota degrade the polysaccharides, and some specific gut bacteria use them as energy source which help in their propagation and release many beneficial compounds mainly short chain fatty acids such as acetate, propionate, butyrate, and valeric acid (Kong et al. 2016; Zhu et al. 2016; Ma et al. 2017). The impacts of fungal polysaccharides on gut microbiota can prove beneficial in the treatment of various health issues such as the autoimmune diseases, allergic disorders, inflammatory bowel disease, irritable bowel syndrome, ulcers, colorectal cancer, necrotizing enterocolitis, and obesity (Milani et al. 2016). Polysaccharides from 53 wild-growing mushrooms have been observed to stimulate *Lactobacillus acidophilus* and *Lactobacillus rhamnosus* growth. These polysaccharides can pass through the stomach unchanged, reach the colon, and enhance the growth of beneficial gut microbes stronger than commercially available prebiotics like inulin or fructooligosaccharides (Nowak et al. 2018). There are different authorities, namely, the World Health Organization (WHO), US Food and Agriculture Organization (FAO), US Food and Drug Administration (FDA), New Zealand Dietary Supplements Regulations (NZDSR), etc., active to regulate the quality, intake, safety, and effectiveness of food and medicinal products including mushroom-derived DSs and functional foods (Wasser and Akavia 2008). To ensure adequate intake of DSs and functional foods, several national medical research bodies such as the Indian Council of Medical Research (ICMR) in India are regularly releasing advisories and recommend the dietary intakes of mushrooms, e.g., *G. lucidum* species (Singh et al. 2020).

References

Adebayo EA, Oloke JK, Aina DA, Bora TC (2014) Antioxidant and nutritional importance of some Pleurotus species. J Microbiol Biotech Food Sci 3(4):289–294

Adedayo MR, Olaschinde IG, Ajayi AA (2010) Nutritional value of some edible mushrooms from Egbe farmland, West Yagba Local government area, Kogi state Nigeria. J Food Sci 4(5):297–299

Akata I, Ergonul PG, Ergonul B, Kalyoncu F (2013) Determination of fatty acid contents of five wild edible mushroom species collected from Anatolia. J Pure Appl Microbiol 7(4):3143–3147

Ao T, Deb CR (2019) Nutritional and antioxidant potential of some wild edible mushrooms of Nagaland, India. J Food Sci Technol 56(2):1084–1089

Azeem U (2017) Taxonomic studies on genus *Phellinus* from district Dehradun (Uttarakhand) and evaluation of some selected taxa for antihyperglycemic activity. Ph.D. Thesis, Department of Botany, Punjabi University, Punjab, India

Badalyan SM (2016) Fatty acid composition of different collections of coprinoid mushrooms (Agaricomycetes) and their nutritional and medicinal values. Int J Med Mushrooms 18(10):883–893

Bandara AR, Rapior S, Mortimer PE, Kakumyan P, Hyde KD, Xu J (2019) A review of the polysaccharide, protein and selected nutrient content of *Auricularia* and their potential pharmacological value. Mycosphere 10(1):579–607

Barreira JCM, Oliveira MBPP, Ferreira ICFR (2014) Development of a novel methodology for the analysis of ergosterol in mushrooms. Food Anal Methods 7(1):217–223

Barros L, Cruz T, Baptista P, Estevinho LM, Ferreira ICFR (2008) Wild and commercial mushrooms as source of nutrients and nutraceuticals. Food Chem Toxicol 46:2742–2747

Bengu AS (2020) The fatty acid composition in some economic and wild edible mushrooms in Turkey. Prog Nutr 22(1):185–192

Borokini P, Lajide L, Olaleye T, Boligon A, Athayde M, Adesina I (2016) Chemical profile and antimicrobial activities of two edible mushrooms (*Termitomyces robustus* and *Lentinus squarrosulus*). J Microbiol Biotech Food Sci 5(5):416–423

Chauhan J, Negi AK, Rajasekaran A, Pala NA (2014) Wild edible macro-fungi – a source of supplementary food in Kinnaur district, Himachal Pradesh, India. J Med Plants Stud 2(1):40–44

Chenghom O, Suksringram J, Morakot N (2010) Mineral composition and germanium contents in some *Phellinus* mushrooms in the Northeast of Thailand. Curr Res Chem 2(2):24–34

Cohen N, Cohen J, Asatiani MD, Varshney VK, Yu H-T, Yang Y-C, Li Y-H, Mau J-L, Wasser SP (2014) Chemical composition and nutritional and medicinal value of fruit bodies and submerged cultured mycelia of culinary-medicinal higher basidiomycetes mushrooms. Int J Med Mushrooms 16(3):273–291

Egwim EC, Elem RC, Egwuche RU (2011) Proximate composition, phytochemical screening and antioxidant activity of ten selected wild edible Nigerian mushrooms. Am J Food Nutr 1(2):89–94

Gaitán-Hernández R, López-Peña D, Esqueda M, Gutiérrez A (2019) Review of bioactive molecules production, biomass, and basidiomata of shiitake culinary-medicinal mushrooms, *Lentinus edodes* (Agaricomycetes). Int J Med Mushrooms 21(9):841–850

Gbolagade J, Ajayi A, Oku I, Wankasi D (2006) Nutritive value of common wild edible mushrooms from Southern Nigeria. Global J Biotechnol Biochem 1(1):16–21

Ghosh K (2016) A review: edible mushrooms as source of dietary fiber and its health effects. J Phys Sci 21:129–137

Giri S, Mandal Subhash C, Acharya K (2013) Proximate analysis of three wild edible mushrooms of West Bengal, India. Int J Pharm Tech Res 5(2):365–369

Jayachandran MJ, Xiao B, Xu (2017) A critical review on health promoting benefits of edible mushrooms through gut microbiota. Int J Mol Sci 18:1934

Jedidi IK, Ayoub IK, Philippe T, Bouzouita N (2017) Chemical composition and nutritional value of three Tunisian wild edible mushrooms. J Food Meas Charact 11:2069–2075

Jiang J, Sliva D (2010) Novel medicinal mushroom blend suppresses growth and invasiveness of human breast cancer cells. Int J Oncol 37(6):1529–1536

Johnsy G, Sargunam SD, Dinesh MG, Kaviyarasan V (2011) Nutritive value of edible wild mushrooms collected from the Western Ghats of Kanyakumari district. Bot Res Int 4(4):69–74

Kolundžić M, Radović J, Tačić A, Nikolić V, Kundaković T (2018) Elemental composition and nutritional value of three edible mushrooms from Serbia. Zastita Materijala 59(1):45–50

Kong Q, Dong S, Gao J, Jang C (2016) In vitro fermentation of sulfated polysaccharides from *E. prolifera* and *L. japonica* by human fecal microbiota. Int J Biol Macromol 91:867–871

Lalotra P, Bala P, Kumar S, Sharma YP (2018) Biochemical characterization of some wild edible mushrooms from Jammu and Kashmir. Proc Natl Acad Sci, India, Sect B Biol Sci 88(2):539–545

Ma G, Kimatu BM, Zhao L, Yang W, Pie F, Hu Q (2017) In vivo fermentation of a *Pleurotus eryngii* polysaccharide and its effects on fecal microbiota composition and immune response. Food Funct 8:1810–1821

Manjula R, Krishnendu A (2012) Proximate composition, free radical scavenging and NOS activation properties of *Ramaria aurea*. Res J Pharmacy Technol 5(11):1421–1427

Milani C, Ferrario C, Turroni F, Duranti S, Mangifesta M, Van Sinderen D, Ventura M (2016) The human gut microbiota and its interactive connections to diet. J Hum Nutr Diet 29:539–546

Musieba F, Okoth S, Mibey RK, Wanjiku S, Moraa K (2013) Proximate composition, amino acids and vitamins profile of *Pleurotus citrinopileatus* singer: an indigenous mushroom in Kenya. Amer J Food Tech 8(3):200–206

Nakalembe I, Kabasa JD, Olila D (2015) Comparative nutrient composition of selected wild edible mushrooms from two agro-ecological zones, Uganda. Springerplus 4(433):1–15

Nowak R, Nowacka-Jechalke N, Juda M, Malm A (2018) The preliminary study of prebiotic potential of Polish wild mushroom polysaccharides: the stimulation effect on *Lactobacillus* strains growth. Eur J Nutr 57(4):1511–1521

Obodai M, Mensah DLN, Fernandes A, Kortei NK, Dzomeku M, Teegarden M, Schwartz SJ, Barros L, Prempeh J, Takli RK, Ferreira ICFR (2017) Chemical characterization and antioxidant potential of wild *Ganoderma* species from Ghana. Molecules 22(196):1–18

Ouzouni PK, Petridis D, Koller W-D, Riganakos KA (2009) Nutritional value and metal content of wild edible mushrooms collected from West Macedonia and Epirus, Greece. Food Chem 115(4):1575–1580

Owaid MN (2018) Bioecology and uses of desert truffles (Pezizales) in the Middle East. Walailak J Sci Tech 15(3):179–188

Pathania J, Chander H (2018) Nutritional qualities and host specificity of most common edible macrofungi of Hamirpur district, Himachal Pradesh. J Biol Chem Chron 4(2):86–89

Perera N, Yang F-L, Lu Y-T, Li L-H, Hua K-F, Wu S-H (2018) *Antrodia cinnamomea* galactomannan elicits immuno-stimulatory activity through Toll-like receptor. Int J Biol Sci 14(10):1378–1388

Phillips KM, Rasor AS (2013) A nutritionally meaningful increase in the vitamin D in retail mushrooms is attainable by exposure to sunlight prior to consumption. Nutr Food Sci 3(6):2155–9600

Phillips KM, Ruggio DM, Horst RL, Minor B, Simon RR, Feeney MJ, Byrdwell WC, Haytowitz DB (2011) Vitamin D and sterol composition of 10 types of mushrooms from retail suppliers in the United States. J Agric Food Chem 59(14):7841–7853

Reis FS, Barros L, Martins A, Ferreiraa ICFR (2012) Chemical composition and nutritional value of the most widely appreciated cultivated mushrooms: an inter-species comparative study. Food Chem Toxicol 50(2):191–197

Reis FS, Martins A, Vasconcelos MH, Morales P, Ferreira ICFR (2017) Functional foods based on extracts or compounds derived from mushrooms. Trends in Food Sci Technol 66:48–62

Rzymski P, Mleczek M, Niedzielski P, Siwulski M, Gąsecka M (2016) Potential of cultivated *Ganoderma lucidum* mushrooms for the production of supplements enriched with essential elements. J Food Sci 81(3):587–592

Safari O, Sarkheil M (2018) Dietary administration of *eryngii* mushroom (*Pleurotus eryngii*) powder on haemato-immunological responses, bactericidal activity of skin mucus and growth performance of koi carp fingerlings (*Cyprinus carpio* koi). Fish Shellfish Immunol 80:505–513

Salamat S, Shahid M, Najeeb J (2017) Proximate analysis and simultaneous mineral profiling of five selected wild commercial mushroom as a potential nutraceutical. Int J Chem Stud 5(3):297–303

Sevindik M, Akgul H, Selamoglu Z, Braidy N (2020) Antioxidant and antigenotoxic potential of *Infundibulicybe geotropa* mushroom collected from Northwestern Turkey. Oxid Med Cell Longev 2020:1–8

Shah N, Usvalampi A, Chaudhary S, Seppänen-Laakso T, Marathe S, Bankar S, Singha R, Shamekh S (2020) An investigation on changes in composition and antioxidant potential of mature and immature summer truffle (*Tuber aestivum*). Eur Food Res Technol 246:723–731

Sharma SK, Gautam N (2017) Chemical and bioactive profiling and biological activities of coral fungi from Northwestern Himalayas. Sci Rep 7:46570–44658

Sinanoglou VJ, Zoumpoulakis P, Heropoulos G, Proestos C, Ćirić A, Petrovic J, Glamoclija J, Sokovic M (2014) Lipid and fatty acid profile of the edible fungus *Laetiporus sulphureus*. antifungal and antibacterial properties. J Food Sci Technol 52(6):1–12

Singh R, Kaur N, Shri R, Singh AP, Dhingra GS (2020) Proximate composition and element contents of selected species of *Ganoderma* with reference to dietary intakes. Environ Monit Assess 192(270):1–15

Singha K, Pati BR, Mondal KC, Mohapatra PKD (2017) Study of nutritional and antibacterial potential of some wild edible mushrooms from Gurguripal Ecoforest, West Bengal, India. Indian J Biotechnol 16:222–227

Sun L, Liu Q, Bao C, Fan J (2017) Comparison of free total amino acid compositions and their functional classifications in 13 wild edible mushrooms. Molecules 22(350):1–10

Thatoi H, Singdevsachan SK (2014) Diversity, nutritional composition and medicinal potential of Indian mushrooms: a review. Afr J Biotechnol 13:523–545

Tietel Z, Masaphy S (2018) True morels (Morchella)—nutritional and phytochemical composition, health benefits and flavor: a review. Crit Rev in Food Sci Nutr 58(11):1888–1901

Toledo CV, Barroetaveña C, Fernandes A, Barros L, Ferreira ICFR (2016) Chemical and antioxidant properties of wild edible mushrooms from native *Nothofagus* spp. forest, Argentina. Molecules 21(1201):1–15

Turfan N, Pekşen A, Kibar B, Ünal S (2018) Determination of nutritional and bioactive properties in some selected wild growing and cultivated mushrooms from turkey. Acta Sci Pol Hortorum Cultus 17(3):57–72

Türkekul I, Çetin F, Elmastaş M (2017) Fatty acid composition and antioxidant capacity of some medicinal mushrooms in Turkey. J Appl Biol Chem 60(1):35–39

Ullah TS, Firdous SS, Mehmood A, Shaheen H, Dar ME (2017) Ethnomycological and nutritional analyses of some wild edible mushrooms from Western Himalayas, Azad Jammu and Kashmir (Pakistan). Int J Med mushrooms 19(10):949–955

Üstün NŞ, Bulam S, Pekşen A (2018) The use of mushrooms and their extracts and compounds in functional foods and nutraceuticals. Türkmen, A. (ed.) 1:1205–1222

Valverde ME, Hernández-Pérez T, Paredes-López O (2015) Edible mushrooms: improving human health and promoting quality life. Int J Microbiol 2015:1–14

Villares A, Mateo-Vivaracho L, Guillamón E (2012) Structural features and healthy properties of polysaccharides occurring in mushrooms. Agriculture 2:452–471

Vishwakarma P, Singh P, Tripathi NN (2016) Nutritional and antioxidant properties of wild edible macrofungi from Northeastern Uttar Pradesh, India. Indian J Tradit Know 15(1):143–148

Wani AB, Bodha RH, Wani AH (2010) Nutritional and medicinal importance of mushrooms. J Med Plant Res 4(24):2598–2603

Wasser SP, Akavia E (2008) Regulatory issues of mushrooms as functional foods and dietary supplements: safety and efficacy. In: Mushrooms as functional foods. Wiley, New York, pp 199–221

Wu D, Pae M, Ren Z, Guo Z, Smith D, Meydani SN (2007) Dietary supplementation with white button mushroom enhances natural killer cell activity in C57BL/6 mice. J Nutr 137(6):1472–1477

Yilmaz F, Işiloğlu M, Merdüvan M (2003) Heavy metal levels in some macrofungi. Turk J Bot 27:45–56

Zhu KX, Nie SP, Tan LH, Li C, Gong DM, Xie MY (2016) A polysaccharide *from Ganoderma atrum* improves liver function in type 2 diabetic rats via antioxidant action and short-chain fatty acids excretion. I J Agric Food Chem 64:1938–1944

Chapter 7
Bioactive Constituents and Pharmacological Activities

The taxa designated as macrofungi represent a wealth of health-promising fungi rich in a wide spectrum of mycochemicals of nutraceutical and therapeutic value (Wasser 2011; Sanico et al. 2014; Rahi and Malik 2016; Hoeksma et al. 2019). However, only a small fraction out of a vast diversity of fungi in millions (nearly 3–5.1 millions) has been explored for nutrients and therapeutic components. In the last few decades, massive research has been carried out and continues to grow incessantly to explore the biochemistry of macrofungi (De Silva et al. 2013; Flores Jr et al. 2014; Dıaz-Godınez 2015; Badalyan 2016; Kalac 2016; Kivrak et al. 2016; Rathore et al. 2017; Elkhateeb et al. 2019a, b). This is necessary in order to give scientific valorization to the traditional practices of macrofungi especially the medicinal uses. The medicinal species largely belong to the genera *Auricularia*, *Cantharellus*, *Ganoderma*, *Pleurotus*, *Lentinus*, *Trametes*, *Tremella*, *Amanita*, etc. and possess a number of high and low molecular weight bioactive constituents responsible for their pharmacological potential (Diksha et al. 2018). Many members of *Basidiomycota* as well as some of the *Ascomycota* are rich in antitumor, antimicrobial, antioxidant, hepatoprotective, hypoglycemic, hypolipidemic, immunomodulating as well as prebiotic substances (Lindequist et al. 2010; Chang and Wasser 2012; Giavasis 2013; Mizuno and Nishitani 2013; Wasser 2014; Singh et al. 2017; Chaturvedi et al. 2018; Fernando et al. 2018; Schüffler 2018; Liu et al. 2020). The major bioactive components present in higher fungi are polysaccharides (homo- and hetero-polysaccharides), terpenes (monoterpenoids, sesquiterpenoids, diterpenoids, and triterpenoids), proteins/peptides, glycoproteins, alkaloids, phenolics, tocopherols, ergosterols, and various fatty acids accounting for their biological activities. These are collectively described as "biological response modifiers" because of their multiple biological impacts which trigger the immune response and elevate the curative properties of the human body (Khatua et al. 2013; Barreira et al. 2014; Kumar 2015; Heleno et al. 2015a; Ruthes et al. 2016; Tel-Cayan et al. 2017; Ma et al. 2018; Phan et al. 2018). The crude extracts and isolated bioactive components from sporocarps and/or mycelial biomass of macrofungi have been appraised for their pharmacological activities. These exhibit anticancer, antidiabetic, anti-inflammatory, antimicrobial, antioxidant, immunomodulatory, hepatoprotective, neuroprotective activities through various metabolic pathways (Thu et al. 2020). However, only a few clinical studies have been performed and need much attention for authentication of the medicinal uses of macrofungi to treat human disorders (Wasser 2017).

Antioxidant Activity

Antioxidants provide protection against oxidative stress resulting from the excessive production of reactive oxygen species (ROS). Several synthetic antioxidants such as butylhydroxyanisole, butylhydroxytoluene, propyl gallate, and tertbutylhydroquinone are in use for years (Atta et al. 2017). However, their safety is argumentative and makes it urgent to find natural sources for novel antioxidants (Ramana et al. 2018). Macrofungi are rich in antioxidants and exhibit free radical scavenging, reducing power, β-carotene bleaching inhibition, thiobarbituric acid reactive substance (TBARS) inhibition properties, etc. (Toledo et al. 2016; Zhang et al. 2018a). Twelve wild strains of *Ganoderma* sp. from Ghana have been found to be rich in fatty acids, sugars (rhamnose, fructose, mannitol, sucrose, and trehalose), β-glucans, organic acids, phenolic compounds, tocopherols, ergosterol, vitamins, and minerals and show antioxidant potential. These constituents display reducing power and play a role in free radical scavenging and lipid peroxidation inhibition (Obodai et al. 2017). Mushroom polysaccharides possess various medicinal properties such as antiaging, antidiabetic, antitumor, antimicrobial, antioxidant, anti-inflammatory, hepatoprotective, neuroprotective, etc. (Ferreira et al. 2015; Ruthes et al. 2016; Li 2017; Kothari et al. 2018; Yu et al. 2018; Yuan et al. 2018). Polysaccharides recovered from *Trametes robiniophila* exhibit appreciable 2,2-azino-bis-3-ethylbenzothiazoline-6-sulfonic acid (ABTS), superoxide anion, and hydroxyl radical scavenging effects in vitro (Wang et al. 2014). The free radical scavenging activities of mushrooms show a dose-dependent trend (Du et al. 2015). Furthermore, the antioxidant potential of mushroom polysaccharides can be improved through the exposure of stress. Heat stress of 42°C for 2h has significantly increased polysaccharide production to nearly 45.63% in *G. lucidum* basidiocarps and also improved antioxidant potential of polysaccharides (Tan et al. 2018). In addition to stress treatment, the structural modifications of polysaccharides, e.g., phosphorylation, elevate the antioxidative efficacy. Phosphorylated polysaccharides of *Russula alutacea* display greater hydroxyl radical, superoxide anion, and 2,2'-azino-bis(3-ethylbenzothiazoline-6-sulphonic acid (DPPH) radical scavenging activity as compared to the unmodified ones (Zhao et al. 2018a). Antioxidant potential of mushrooms makes immune health strong. The *S. granulatus* polysaccharide (SGP) fractions SGP I-b and SGP II-b, purified from the basidiocarps of *Suillus granulatus* enhance lymphocyte proliferation in vitro through free radical scavenging and reduction capacity (Chen et al. 2018a, b). *P. involutus* polysaccharides through their scavenging ABTS, DPPH, and hydroxyl and superoxide radicals modulate the immune response increasing release of TNF-α and interleukin-6 (IL-6) from RAW264.7 cells (Liu et al. 2018). A variety of phenolics such as phenolic acids, flavonoids, hydroxybenzoic acids, hydroxycinnamic acids, lignans, tannins, stilbenes, and oxidized polyphenols are also present in macrofungi (Palacios et al. 2011). They form a class of bioactive mycochemicals with significant free radical scavenging and ferric reducing antioxidant power (FRAP) (Yahia et al. 2017). Several yellow polyphenol pigments referred to as styrylpyrones are found in medicinal mushrooms, e.g., *Phellinus* and *Inonotus* species (Hymenochaetaceae) with plausible antioxidant potential. Coculturing like that of *I. obliquus* with *Phellinus punctatus* might prove an inexpensive strategy for upregulating the biosynthesis of bioactive compounds. Enhanced production and variations in metabolic profiles involving various metabolites such as inoscavin B (1), inoscavin C (2), methylinoscavin A (3), davallialactone (4), and methyldavallialactone (5) (Fig. 7.1; 1–5) from dual cultures of different mushrooms could prove helpful in the discovery of new bioactive compounds (Zheng et al. 2011). In addition to contributing towards bitterness of mushrooms such as *Pholiota* and *Hypoholoma*, styrylpyrone pigments have role in various pharmacological activities such as anticancer, antidiabetic, antioxidative, antitumor, and antiviral and are of great significance in various pharmaceutical applications (Lee et al. 2008; Lee and Yun 2011; Ayala-Zavala et al. 2012).

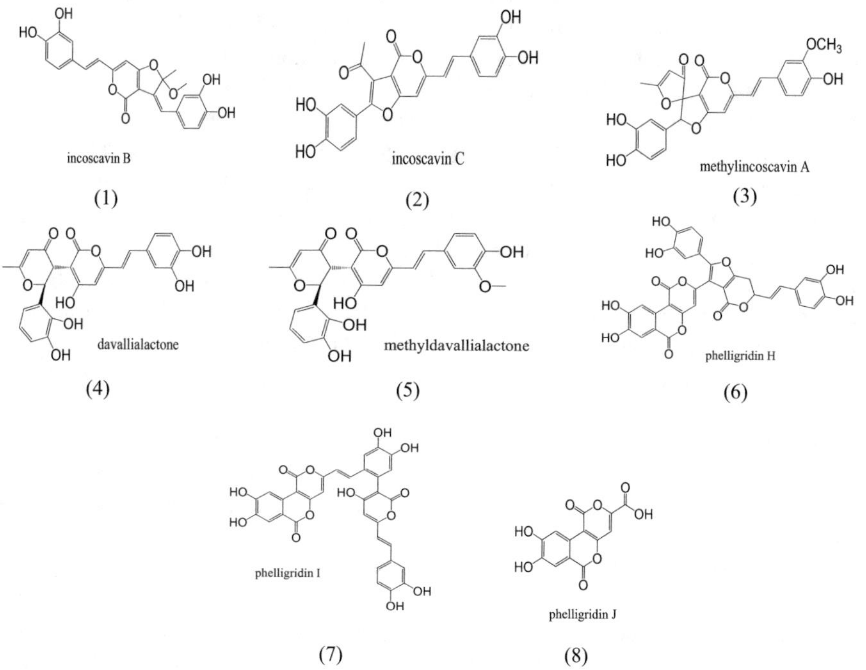

Fig. 7.1 Chemical structures of selected styrylpyrone derivatives in the genera *Phellinus* and *Inonotus* (*Hymenochaetaceae*)

The styrylpyrone pigments, e.g., inoscavins A–E, methylinoscavins A–D, inonoblins A–C, phelligridins D–E, and G, isolated from the basidiocarps of *Phellinus* and *Inonotus* have the ability of free radical scavenging against ABTS, DPPH, and superoxide radicals and to prevent lipid peroxidation (Wang et al. 2005; Lee and Yun 2007; Lee et al. 2007a,b; Jung et al. 2008). A number of biological activities have been credited to the species classified in the genus *Phellinus* because of the presence of various bioactive constituents including styrylpyrones (Azeem et al. 2018). The phelligridins H–J (Fig.7.1; 6–8) of *Ph. igniarius* exhibit antioxidant and cytotoxic potential (Wang et al. 2007a; Lu et al. 2009; Huang et al. 2010). Antioxidant properties of hispidin defend from inflammation through blockade of nuclear factor kappa B (NF-$_k$B) activation in mouse macrophage cells (Shao et al. 2015). Hispolon, another phenolic antioxidant compound isolated from *Ph. linteus*, decreases the level of malondialdehyde (MDA) in the edema paw and enhances the activities of superoxide dismutase (SOD), glutathione peroxidase (GPx), and glutathione reductase (GRx) in the liver (Chang et al. 2011). *Macrocybe lobayensis* is an edible mushroom and due to its aroma and taste becomes an integral part of tribal cuisines of India. This mushroom contains phenolic compounds (p-hydroxybenzoic acid, p-Coumaric acid, cinnamic acid, salicylic acid, pyrogallol) as well as ascorbic acid and carotenoids which probably contribute towards its antioxidant potential evident from the ABTS and DPPH radical scavenging, chelating ability of ferrous ion, reducing power, and total antioxidant capacity (Khatua et al. 2017). The phenol-rich extract of wood ear mushroom, *A. nigricans*, cultivated on rubberwood sawdust causes DPPH; hydroxyl, superoxide anion; and peroxyl radical scavenging and exhibits reducing power (Teoh et al. 2019). Three polyketide-type antioxidant compounds, cyathusals A–C, and pulvinatal have been obtained from the cultures of *Cyathus stercoreus* (the coprophilous mushroom). These exhibit free radical scavenging efficiency against DPPH radical with IC$_{50s}$ of 41.6, 46.0, 26.6, and 28.6 µM, respectively, and on the ABTS radical with IC$_{50s}$ of 7.9, 11.1, 9.1, and 8.4 µM, respectively (Kang et al. 2007). Total phenols along with metal nutrients have been detected in edible mushrooms collected from Sinop (Turkey), namely, *A. caesarea*, *B. edulis*, *G. frondosa*, *H. repandum*, *L. deliciosus*, *L. piperatus*, *L. volemus*, *L. sulphureus*, *P. ostreatus*, and *R. flava*. These species exhibit antioxidant properties (free radical scavenging activity, lipid peroxidation inhibition, and metal chelation ability) and hence might prove as promising DSs to prevent various infectious diseases (Ozen et al. 2019). *Tricholoma populinum*, *T. scalpturatum*, *Neolentinus cyathiformis*, *Chlorophyllum agaricoides*, and *Calvatia utriformis* are traditional foods in Turkey for a long time. Chromatography of these mushroom extracts shows the presence of phenolic acids (protocatechuic, gallic, and chlorogenic acids) and fatty acids (oleic, linoleic, and palmitic acids) responsible for their antioxidant activities (Sezgin et al. 2020). *P. cystidiosus* mushroom has been found to be rich in 11 mycochemicals including essential oil, triterpenes, anthraquinones, tannins, flavonoids, phenols, fatty acids, alkaloids, steroids, sugars, and coumarins. Furthermore, ethanolic extract of mycelial strain (WS218-2) of *P. cystidiosus* have shown the highest radical scavenging activity (79.01%) with maximum phenolic content (95.67 mg gallic acid equivalents/g of sample) (Garcia et al. 2020). Two samples of

M. esculenta, one of the most highly prized edible mushrooms collected from Portugal and Serbia, have been found to be rich in carbohydrates (including free sugars), proteins, saturated fatty acids, monounsaturated and polyunsaturated fatty acids, and many other bioactive compounds (organic acids, phenolic compounds, and tocopherols). These mushrooms display antioxidant properties (radical scavenging, reducing power, and lipid peroxidation inhibition) in addition to antimicrobial activity (antibacterial and demelanizing properties) (Heleno et al. 2013). The polyunsaturated fatty acids, ergosterol, and tocopherols are scavengers of free radicals and protect from the hazardous effects of ROS in degenerative, microbial, and cardiovascular diseases (Guillamón et al. 2010; Heleno et al. 2012; Kalac 2013; Heleno et al. 2015b; Jaworska et al. 2015).

Antitumor Activity

Human beings are continuously suffering from the scourge of cancer, a disease characterized by uncontrolled cell growth. It has been reported to be one of the leading causes of death globally after cardiovascular disorders accounting for an estimated 9.6 million deaths or 1 in 6 deaths in 2018 (https://www.who.int/news-room/fact-sheets/detail/cancer accessed on May16, 2020). Fungi are new attractive sources of natural antitumor compounds for future cancer therapy (Blagodatski et al. 2018). Polysaccharides from different natural sources such as plants, animals, microbes, and fungi have broad application prospects including anticancer efficacy (Ling and Gangliang 2018). They when obtained from *P. ostreatus* have been observed to reduce the colony size and invasive capabilities of BGC-823 cells (Cao et al. 2015). The polysaccharides extracted from *P. eryngii* are able to arrest the cell cycle at the S-phase, stimulating the production of ROS in HepG-2 cells, suppressing their proliferation, and increasing the exit of lactate dehydrogenase from HepG-2 cells (Ma et al. 2014a, b; Ren et al. 2016). Polysaccharides from *G. lucidum* activate the mitogen-activated protein kinase (MAPK) pathways in human leukemia (HL-60) cells (Yang et al. 2016). Antitumor potential of *A. bisporus* can be credited to polysaccharides inducing apoptosis through mitochondrial death pathway (Pires et al. 2017). *L. edodes* polysaccharides induce apoptosis of human colon cancer cells through ROS-mediated intrinsic and TNF-α-mediated extrinsic pathway (Wang et al. 2017). Proteins provide multiple health benefits such as improvement of digestion and absorption of exogenous nutritional components stimulating the immune system and defending the host from the entry of pathogens and inhibition of some enzyme activities (Petrovska 2001). Overall, proteins and peptides have pharmaceutical significance such as lectins, fungal immunomodulatory proteins, ribosome-inactivating proteins, ribonucleases, and laccases (Xu et al. 2011). Fungal immunomodulatory proteins possess antitumor potential as they inhibit the invasion and metastasis of tumor cells (Lin et al. 2010). *G. tsugae* and *G. microsporum* contain fungal immunomodulatory proteins, which cause growth inhibition of cancer cells, and induce autophagy-dependent caspase-independent cell death through the accumulation of autophagosome (Hsin et al. 2015, 2016). Lectins obtained from *M. procera* show antitumor activities by binding two glycoproteins (aminopeptidase N (CD13), integrin $\alpha 3\beta 1$) that are overexpressed on the membrane of tumor cells and allow the entry of protein drugs into cancer cells and *Ganoderma lipsiense* against human tumor (HT-29) colon adenocarcinoma cells (Kumaran et al. 2017; Zurga et al. 2017). *G. lucidum, Lyophyllum shimeji*, and *Tuber indicum* ribonucleases possess antitumor potential against hepatoma (HepG2), human breast cancer cell lines (MCF7), and HCT116 cells (Zhang et al. 2010; Dan et al. 2016; Xiao et al. 2017). *G. lucidum, Sanghuangporus baumii, C. comatus*, and *Cerrena unicolor* laccases show antitumor activities against HepG2 and L1210 cells and leukemic cells (Matuszewska et al. 2016). The low molecular weight subfractions isolated from the secondary metabolites of white rot fungus, *C. unicolor*, exhibit antioxidant and antitumor activities in human colon cancer cells (stage 1) HT-29. These subfractions result in apoptosis and suppress the proliferation of cells (47.5–9.2%) at the maximum concentration, and this effect exhibits a parallel trend with the dose (Matuszewska et al. 2019). Several terpenes including sesquiterpenoids (aristolane, bisabolane, cuparene, drimane, fomannosane, lactarane, nordasi-nane, spiro, sterpurane), diterpenoids of cyathane type, and triterpenoid of lanostane type are present in macrofungi (Duru and Cayan 2015). Irofulven, named as 6-hydroxymethylacylfulvene and MGI-114 derived from illudin-S, a sesquiterpenoid obtained from *O. olearius* cultures have antitumor effects (McMorris et al. 1996; McMorris 1999; Schobert et al. 2011). Irofulven inhibits synthesis of DNA and stimulates apoptosis even in the nM range (Woynarowski et al. 1997; Kelner et al. 2008). The flammulinol A (9), an isolactarane sesquiterpene (10), seven isolactarane-related norsesquiterpenes, flammulinolides A–G (11–17), and sterpuric acid have been obtained from *F. velutipes* culture (Fig. 7.2). These compounds exhibit cytotoxic effects on KB cells with IC_{50s} of 3.6–4.7 µM. Flammulinolide C exerts cytotoxic action on HeLa cells with IC_{50} of 3.0 µM (Wang et al. 2012a), and those obtained from *Inonotus rickii* act against human colon cancer SW480 (Chen et al. 2014). From the fermentation products of *Coprinellus radians*, 13 new guanacastane-type diterpenoids, named radianspenes, have been isolated, and these terpenes exert antitumor effects against MDA-MB-435 cells (Ou et al. 2012). Diterpenoids present in *P. eryngii* produce cytotoxic effects on human cancer cell lines (Wang et al. 2012b). Triterpenoids from *G. lucidum* exhibit excellent antitumor potential. Ganoderic acid T (18) causes apoptosis of metastatic lung tumor cells through an intrinsic pathway linked to mitochondrial dysfunction (Tang et al. 2006a). A triterpenoid, ganoderic acid DM (19) recovered from *G. lucidum* shows antiproliferation activity and effectively suppresses MCF-7 human breast cancer cell proliferation. It is believed to induce cell cycle (G1) arrest and apoptosis in MCF7 cells (Liu et al. 2012a; Wu et al. 2012). Furthermore, lucidenic acids A–C (20–22) and N (23) obtained from the basidiocarps of a new strain of *G. lucidum* (YK-02) extracts exhibit anti-invasive effects on hepatoma cells (Weng et al. 2007). Additionally, ganoderic acid named as 3α, 22β-diacetoxy-7α-hydroxy-5α--lanosta-8,24E-dien-26-oic acid (24) (Fig. 7.3; 18–24) identified from *G. lucidum* mycelia has displayed considerable cytotoxicity (Li et al. 2013). Aside from other bioactive metabolites, lanostane-type triterpenoids have also been isolated from

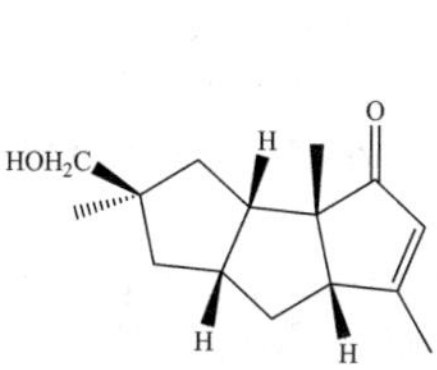

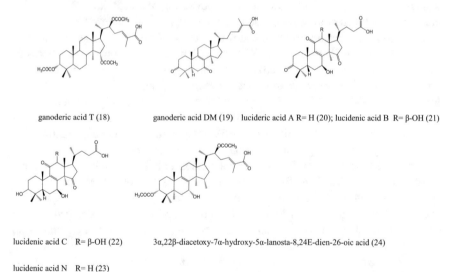

Fig. 7.2 Chemical structures of cytotoxic terpenoids from *Flammulina velutipes*

flammulinol (9)

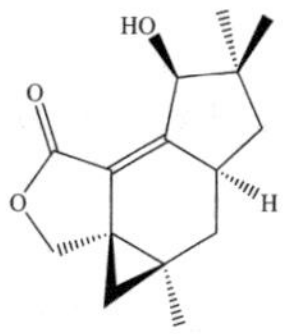

flammulinolide A (10)

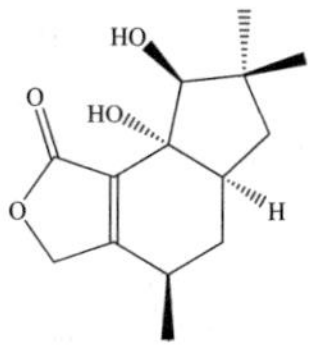

flammulinolide B (11)

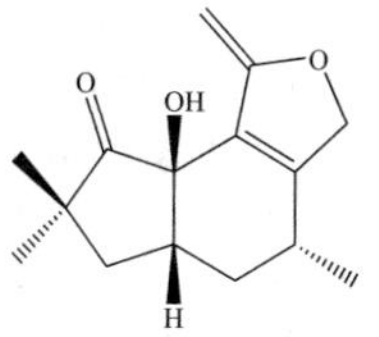

flammulinolide C (12)

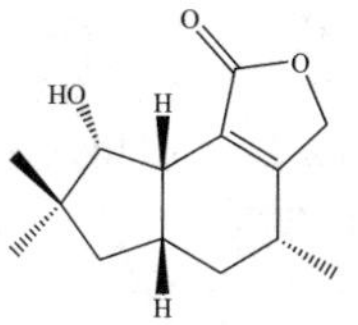

flammulinolide D (13)

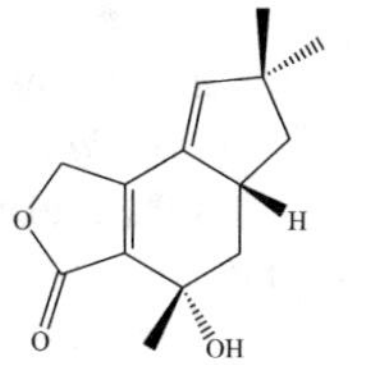

flammulinolide E (14)

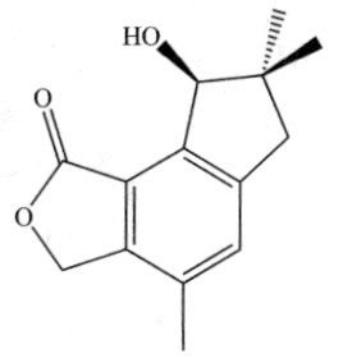

flammulinolide F (15)

ganoderic acid T (18)

ganoderic acid DM (19)

lucideric acid A R= H (20); lucidenic acid B R= β-OH (21)

lucidenic acid C   R= β-OH (22)

lucidenic acid N   R= H (23)

3α,22β-diacetoxy-7α-hydroxy-5α-lanosta-8,24E-dien-26-oic acid (24)

Fig. 7.3 Chemical structures of antitumor triterpenoids from *Ganoderma lucidum*

I. obliquus with potent antitumor efficacies (Zheng et al. 2010; Kim et al. 2011). Earlier studies revealed the structures of these compounds isolated from the sclerotia of this mushroom: inonotsuoxides A and B (Nakata et al. 2007); inonotsulides A, B, and C (Taji et al. 2007); inonotsutriols A–C (Taji et al. 2008a); inonotsutriols D–E (Tanaka et al. 2011); and lanosta-8,23E-diene-3β,22R,25-triol, lanosta-7:9(11),23E-triene-3β,22R,25-triol, and 3β-hydroxylanosta-8,24-dien-21-al

(Taji et al. 2008b). Inotodiol isolated from the same mushroom exhibits antitumor activity through caspase-3-dependent apoptosis in tumor cells (Nomura et al. 2008). Other lanostane-type triterpenes obtained from sclerotia of *I. obliquus* are spiroino-notsuoxodiol, lanostane-type triterpenoids, inonotsudiol A, and inonotsuoxodiol A with cytotoxic effects (Handa et al. 2010). Lanostanes (3β,15α-dihydroxylanosta-7,9(11),24-triene-21-oic acid, dehydroeburicoic acid, 15α-acetyl-dehydrosulphurenic acid, dehydrosulphurenic acid and sulphurenic acid) and ergostane-type triterpenes (methyl zhankuic acid A, zhankuic acid A, and zhankuic acid C isolated from basidiocarps of *T. camphoratus*) exhibit in vitro cytotoxicity against various cancer cell lines including human breast cancer cells (Yeh et al. 2009). The lanostane-type triterpenoid, 25-methoxyporicoic acid A, isolated from the epidermis of the sclerotia of *Wolfiporia extensa* suppresses skin tumor promotion under in vivo investigations (Akihisa et al. 2009). From ethyl acetate crude extract of *Ganoderma zonatum*, lanostane-type triterpenoids, steroids, and a benzene derivative have been isolated. Lanostane triterpenoid, ganoderic acid Y shows cytotoxicity against two human cancer cell lines, SMMC-7721 (liver cancer) and A549 (lung cancer) (Kinge and Mih 2011). Antrocin obtained from the basidiocarps of *T. camphoratus* has anticancer activity against MDA-MB-231 and MCF-7 cells with an IC_{50} value of 0.6 µM (Rao et al. 2011). The bioactive compound 4,7-dimet hoxy-5-methyl-1,3-benzodioxole isolated from the basidiocarps of *T. camphoratus* displays potent in vivo antitumor effects by activating p53-mediated p27/Kip1 signaling pathway (Tu et al. 2012). A lanostane triterpene glycoside, namely, fomitoside-K, from *Fomitopsis nigra* activates apoptosis of human oral squamous cell carcinomas (YD-10B) through the ROS-dependent mitochondrial dysfunction signaling pathway (Bhattarai et al. 2012; Lee et al. 2012). Triterpenoids obtained from *Hypholoma fasciculare* show anticancer properties against four human cancer cell lines (A549, SK-OV-3, SK-MEL-2, and HCT-15) (Kim et al. 2013). Further, another class of terpenoids, sesterpenoids (C-25), displays a number of bioactivities including anticancer activity (Antonio et al. 2015). Styrylpyrone compounds phelligridins C–F cause cytotoxicity against a human lung cancer cell line and a liver cancer cell line, and phelligridins C and D show potent anticancer effects with IC_{50s} in the range of 100 nM against A549 and Bel7402 (Mo et al. 2004). Hispolon found in *Phellinus* species has been observed to induce epidermoid and gastric cancer cell apoptosis and regardless of p53 status inhibited the breast and bladder cancer cell growth (Patel and Goyal 2012). The methanol extracts of *Agaricus arvensis, A. urinascens, Albatrellus pes-caprae, A. caesarea, A. citrine, A. rubescens, Boletus aereus, B. edulis, C. gambosa, C. cibarius, C. cinereus, Craterellus cornucopioides, F. velutipes, G. lucidum, H. repandum, Hygrophorus marzuolus, H. russula, I. geotropa, L. volemus, M. mastoidea, M. conica, M. elata, M. esculenta, M. procera, P. eryngii, P. ostreatus, R. cyanoxantha,* and *R. virescens* species, rich in phenols and flavonoids, possess antioxidant, antitumor, antiproliferative, cytotoxic, and proapoptotic activities against human lung adenocarcinoma cell line A549 by the MTT. Piceatannol (a phenolic compound) has been found as the bioactive compound responsible for the antiproliferative action of these species (Vasdekis et al. 2018).

Antidiabetic Activity

Diabetes mellitus is a chronic carbohydrate metabolic syndrome leading to several health problems related to hyperglycemia and occurs in people in low- to middle-income countries. The population of diabetics has risen from 108 million in 1980 to 422 million in 2014 according to WHO estimates (https://www.who.int/news-room/fact-sheets/detail/diabetes accessed on May16, 2020). The major cause of diabetes mellitus is insufficient production of insulin and/or insulin resistance and is diagnosed by persistent high concentration of glucose in the blood (Zaccardi et al. 2016). Edible macrofungi represent ideal diets for diabetics and low in energy with negligible amount of fats and cholesterol and contain carbohydrates, fibers, high content of proteins, vitamins, and minerals (Cheung 2013; Dubey et al. 2019). The sporocarps and mycelia of macrofungi either in powder form or extracts made in different solvents exhibit antidiabetic efficacies through different metabolic pathways resulting in decline in blood glucose level, elevation in insulin level and sensitivity, lowering of cholesterol and triglyceride levels, etc. (Kaur et al. 2015; Srivastava et al. 2018; Gulati et al. 2019). Alpha-glucosidase represents the most important enzyme involved in carbohydrate digestion inside the human body. Therefore, inhibiting the activity of this enzyme may prove beneficial in the treatment of type 2 diabetes mellitus. The hydroalcoholic extracts of *Phellinus* species enriched with different mycochemicals display antidiabetic potential through the inhibition of carbohydrate-digesting enzymes, α-amylase and α-glucosidase (Azeem 2017). An acid protein-bound polysaccharide derived from *I. obliquus* displays an inhibitory effect on α-glucosidase (an enzyme-cleaving carbohydrates to glucose) with inhibitory concentration (IC_{50}) of 93.3 μg/mL and prevents lipid peroxidation in rat liver tissues (Chen et al. 2010). A lanostane triterpenoid from *G. lucidum*, namely, ganoderol B-[(3β,24E)-lanosta-7,9(11),24-trien-3, 26-diol], has strong inhibitory action on α-glucosidase (Fatmawati et al. 2011). The ethanol extracts of *Phellinus merrillii* containing hispidin (25), hispolon (26), and inotilone (27) (Fig. 7.4; 25–27) display significant α-glucosidase and aldose reductase inhibitory activities in vivo (Lee et al. 2010; Huang et al. 2011). All the three compounds, namely, hispidin, hispolon, and inotilone, have shown inhibition of α-glucosidase with IC_{50}s of 297.06, 12.38, and 18.62 μg/mL, respectively, while aldose reductase inhibition with IC_{50} of 48.26, 9.47, and 15.37 μg/mL, respectively. In addition, hispidin from *Ph. linteus* indirectly exerts antihyperglycemic effect through β-cell protection from the toxic actions of ROS during diabetes (Jang et al. 2010). *P. ostreatus*, *C. indica*, and *V. volvacea* cause concentration-dependent inhibition of α-amylase and increase in glucose transport across yeast cells in vitro. *C. indica* extracts have shown maximum enzyme inhibition and glucose uptake by yeast cells in vitro and at 400 mg/kg dose lower blood glucose concentration comparable to acarbose by increasing starch tolerance in starch-fed mice (Singh et al. 2017). In addition to the inhibition of carbohydrate-digesting enzymes, mycochemicals follow various other metabolic routes to combat diabetes mellitus. *P. eryngii* polysaccharides increase the level of high-density lipoprotein cholesterol and liver glycogen, while *P. ostreatus*

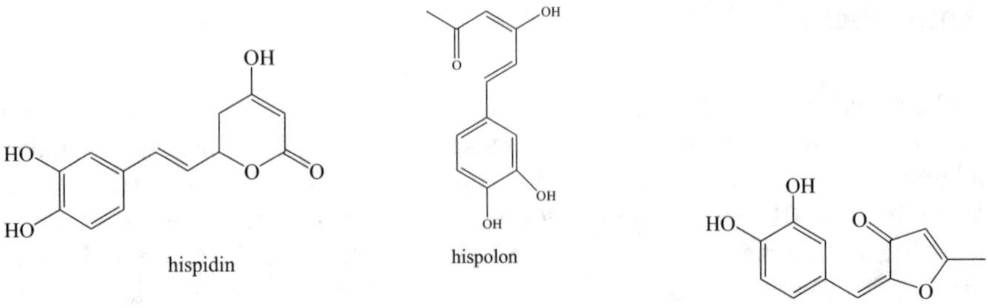

hispidin   hispolon

(25)   (26)   (27)

Fig. 7.4 Chemical structures of bioactive compounds hispidin, hispolon, and inotilone

polysaccharide-rich extracts activate GSK3 phosphorylation and GLUT4 transloca-
tion in streptozotocin (STZ)-induced diabetic rats (Chen et al. 2016; Zhang et al.
2016). *G. lucidum* polysaccharides show antidiabetic activity by downregulation of
hepatic glucose-regulated enzyme mRNA levels through AMPK activation, lowering of epididymal fat/BW ratio, elevation of insulin sensitivity, and improvement of
gut microbiota composition (Xu et al. 2017). Three polysaccharides extracted from
S. luridus (Suilu-A, Suilu-C, and Suilu-S) have shown potential antidiabetic effects
and improvement in pathologic morphologies of the liver and kidney in STZ-
induced diabetic mice (Zhang et al. 2018b). Another strategy to combat hyperglycemia is through the inhibitory effect on the formation of advanced glycation end
products. *A. auricula-judae* polysaccharides and their hydrolysates significantly
inhibit the formation of advanced glycation end products in short- and long-term
glycosylation in a dose-dependent manner and extend the lifespan of *Caenorhabditis
elegans* cells by 32.9% under high sugar stress (Shen et al. 2019). Hyperglycemia
exacerbates ROS production leading to mitochondrial dysfunction. ROS production
arises from the NADPH oxidases and/or from the mitochondria in hyperglycemic
cells (Dymkowska et al. 2014). Excessive glucose entering into the susceptible diabetic cells, such as endothelial cells, would cause hike in mitochondrial pyruvate
and reduced substrates, NADH and FADH2, in the electron transport chain (ETC)
leading to ETC dysfunctioning (Nishikawa et al. 2000). This may cause production
of superoxide anion radicals in mitochondrial complex I (NADH: ubiquinone oxidoreductase) and III (ubiquinol: cytochrome c oxidoreductase) (Halliwell 2006).
The excess of ROS causes lipid, protein, and DNA damage, cell dysfunction, and
apoptosis adding to diabetic complications (Dymkowska et al. 2014; Hyvönen et al.
2015; Rahvar et al. 2017). *Pleurotus albidus* extract inhibits the increase in the
activity of complex I of the electron transport chain and reduces ROS production
induced by hyperglycemia in EA.hy926 endothelial cells. Moreover, this extract
prevents oxidative damage to lipids and proteins, modulates activities of SOD and
catalase (CAT), and lowers the nitric oxide production under hyperglycemia
(Gambato et al. 2018). *A. bisporus*, an edible mushroom cultivated in the world
widely, has been found to be rich in flavonoids, alkaloids, terpenoids, and saponins.

The hydroalcoholic extract of *A. bisporus* has antidiabetic potential bringing decrease in blood glucose through the amelioration of oxidative stress, i.e., lowering MDA levels and increasing SOD activity in alloxan-induced diabetic rats (Ekowati et al. 2018). The ethanolic and aqueous extracts of *Ganoderma cupreolaccatum* and *G. resinaceum* enriched with proteins, phenols, and flavonoids exhibit antidiabetic activity. The said extracts show antidiabetic action by increasing body weight, elevating glucose tolerance of rats subjected to oral glucose tolerance test (OGTT), bringing changes in pancreatic and liver morphology, and increasing the antioxidant potential in alloxan-induced diabetic rats (Raseta et al. 2020).

Antibacterial and Antifungal Activities

A number of infectious diseases caused by bacteria, viruses, and fungi or other parasites are among the major causes of morbidity and mortality throughout the world. The most serious problem in the treatment of such diseases is the developing multiresistant nature of the causal agents. This emphasizes the need to explore new drug resources against these microbes (Prestinaci et al. 2015; Vacca et al. 2018; Watkins et al. 2019). Macrofungi possess antibacterial and antifungal components such as polysaccharides, terpenoids, phenolics, proteins, etc. which may prove beneficial in the discovery of novel antimicrobial drugs to combat microbial infections (Alves et al. 2012; Bal 2019; Gebreyohannes et al. 2019). Extracellular and intracellular polysaccharides from submerged culture of *Cordyceps cicadae* inhibit the growth of *Escherichia coli*, *Klebsiella pneumoniae*, *Vibrio alginolyticus*, *V. cholerae*, *V. parahaemolyticus*, *Pseudomonas aeruginosa*, *Staphylococcus aureus*, and *Streptococcus pneumoniae* (Sharma et al. 2015). Hot water extract, partially purified polysaccharides, and hot alkali extract of wild mushroom *G. lipsiense* exhibit antimicrobial potential against five Gram-negative (*Proteus hauseri*, *E. coli*, *Salmonella enteritidis*, *Shigella sonnei*, and *Yersinia enterocolitica*) and five Gram-positive (*Listeria monocytogenes*, *S. aureus*, *B. cereus*, *Geobacillus stearothermophilus*, and *Enterococcus faecalis*) bacterial strains and two fungal strains (*Candida albicans* and *Cryptococcus neoformans*) (Klaus et al. 2017). Antimicrobial activities of several other proteins/peptides of fungal origin have been documented in literature. For example, lectins from *Sparassis latifolia* inhibit growth of *E. coli*, *P. aeruginosa*, and resistant strains of *S. aureus* and fungal species, namely, *Candida* and *Fusarium* spp. (Chandrasekaran et al. 2016). Fungal ribonucleases have remarkable antibacterial activity against *P. aeruginosa*, *P. fluorescens*, and *S. aureus* at the RNA level (Alves et al. 2013). Plectasin is a cysteine-rich host defense peptide and is the first defensin extracted from *Pseudoplectania nigrella* that displays significant antibacterial efficacy against *S. pneumoniae* (Mygind et al. 2005). Hybrid antimicrobial peptide magainin II-cecropin B (Mag II-CB) isolated from the medicinal fungus *C. militaris* exhibits antibacterial and immunomodulatory effects in BALB/c mice infected with *E. coli*. Mag II-CB has been found to ameliorate *E. coli*-related intestinal damage, upregulate tight junction proteins (zonula occludens-1, clau-din-1, and occludin), positively modulate the intestinal microbial flora, and regulate plasma immunoglobulin and ileum secreted immunoglobulin A levels by attenuat-ing their pro-inflammatory cytokine levels and elevating their anti-inflammatory cytokines levels in mice infected with *E. coli* (Zhang et al. 2018c). *S. aureus* infec-tions cause illness, mortality, and serious economic burden on patients across the world and require novel antibacterial drugs urgently to treat these infections (Klein et al. 2007; Boucher and Corey 2008). Macrofungal species rich in anti-*Staphylo-coccus* constituents might prove as a natural source for new drug discovery against *S. aureus* infections. Coprinol, an antibacterial cuparane-type terpenoid obtained from the cultures of a *Coprinus* sp., shows anti-*Staphylococcus* effect along with bactericidal potential against multidrug-resistant Gram-positive bacteria, namely, *Escherichia coli*, *Pseudomonas* sp., *Haemophilus influenzae*, *S. pneumoniae*, and *Enterococcus faecium* (Johansson et al. 2001). The sesquiterpenoid-rich ethyl ace-tate extract of *G. praelongum* shows maximum antibacterial activity at 35.67 ± 0.62 μM with minimum inhibitory concentration (MIC) of 0.390–6.25 mg/ mL (Ameri et al. 2011). In addition to their inhibitory action on the growth of *E. coli* and *Bacillus subtilis*, sesquiterpenoids from *F. velutipes* also have the potential to inhibit the growth of methicillin-resistant *S. aureus* and can prove helpful in treating the infections of *S. aureus* (Wang et al. 2012a). *G. lucidum*, *G. praelongum*, and *G. resinaceum* have been tested against 30 strains of clinical isolates of methicillin-resistant and methicillin-sensitive *S. aureus*. *Armillaria* species have great antimi-crobial potential credited to several sesquiterpene aryl esters present in them. Arnamial from *A. mellea* proves to be the most effective compound against *Penicillium oxalicum* and four basidiomycetes with MICs of <5 μg/mL (Misiek and Hoffmeister 2012). The sesquiterpenoids, namely, udasterpurenol A (28) and uda-lactaranes A (29) and B (30) (Fig. 7.5; 28–30), have been obtained from *Mycoacia uda*. The latter have the potential to inhibit the germination of *Fusarium*

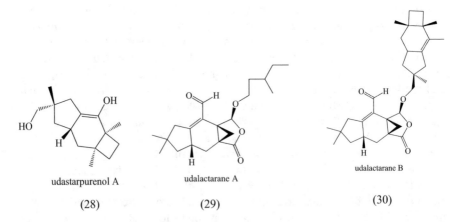

udastarpurenol A

(28)

udalactarane A

(29)

udalactarane B

(30)

Fig. 7.5 Chemical structures of antifungal sesquiterpenoids from *Mycoacia uda*

graminearum spores, a plant pathogenic fungus (Schüffler et al. 2012). A diterpenoid, pleuromutilin, is a well-known antibiotic derived from the fungus *C. passeckerianus* (Novak and Shlaes 2010). This compound has resulted in the discovery of an improved antibiotic, retapamulin, which is a C14-sulfanyl-acetate derivative of pleuromutilin and later on developed into an antibiotic drug (Nagabushan 2010). In order to get rid of the side effects of current medicines and to reduce the multidrug resistance of *Mycobacterium tuberculosis*, antitubercular compounds, particularly those from natural products, have gained tremendous research interest (Ginsberg 2010; Barrios-Garcia et al. 2012). Lanostane triterpenes, astraodoric acid A from *Astraeus pteridis* and astraodoric acid B from *A. odoratus*, have been observed with MICs of 50 and 25 µg/mL against *M. tuberculosis*, respectively (Stanikunaite et al. 2008; Arpha et al. 2012). Butenolides and ramariolides A–D have been isolated from the basidiocarps of the coral mushroom, *R. cystidiophora*. These compounds exhibit in vitro antibacterial effects against *Mycobacterium smegmatis* and *M. tuberculosis* (Centko et al. 2012). Moreover, lanostane triterpenoids, ganorbiformins A–G, have been reported from *Ganoderma orbiforme* BCC 22324, and the C-3 epimer of ganoderic acid T possesses significant antimycobacterial activity with MIC 1.3 µM (Isaka et al. 2013). Many more lanostane triterpenoids obtained from *Albatrellus flettii, F. pinicola, F. rosea*, and *Jahnoporus hirtus* displayed anti-*Bacillus* and anti-*Enterococcus* activities (Popova et al. 2009; Liu et al. 2010a, b). Polyphenols exhibit excellent antimicrobial potential (Othman et al. 2019). Phenol- and flavonoid-rich extracts of *B. aestivalis, B. edulis*, and *L. pseudoscabrum* mushrooms exhibit significant antioxidant activity; bactericidal potential against *S. aureus, E. coli, K. pneumoniae, P. aeruginosa*, and *Enterococcus faecalis*; and antifungal efficacies against *Aspergillus flavus, A. fumigatus, Candida albicans, Paecilomyces variotii*, and *Penicillium purpurescens* (Kosanić et al. 2019). Six polyphenolics, e.g., rutin, chlorogenic acid, quercitrin, isorhamnetin, quercetin, and icariside II, have been identified in the ethyl acetate fraction of ethanol extract prepared from *Sanghuangporus sanghuang* basidiocarps. This ethyl acetate fraction exhibits remarkable antioxidant, antihyperglycemic, and antimicrobial potential (Liu et al. 2017). *Fatty acid-rich petroleum ether extract of Pleurotus eous basidiocarps shows significant bactericidal activity against Bacillus cereus, B. subtilis, E. coli, K. pneumoniae, P. aeruginosa*, and *S. aureus with MICs of* 4.2 µg/mL, 3.1 µg/mL, 3.1 µg/mL, 4.4 µg/mL, 8.8 µg/mL, and 4.4 µg/mL, *respectively (*Suseem and Saral 2013). The extracts made from the basidiocarps of *L. sulphureus* (a saprophyte growing on deciduous trees) have been found to be rich in different classes of fatty acids. These extracts have been evaluated for bactericidal activity against Gram-negative bacteria, *E coli, P. aeruginosa*, and *Salmonella typhimurium*; Gram-positive bacteria, *Listeria monocytogenes, Micrococcus flavus*, and *S. aureus*; and micromycetes, *Aspergillus fumigatus, A. versicolor, A. ochraceus, Trichoderma viride, Penicillium ochrochloron*, and *P. verrucosum* var. *cyclopium* (Sinanoglou et al. 2014).

Antimalarial Activity

Malaria is among the most devastating parasitic infectious diseases in many sub-tropical and tropical regions caused by *Plasmodium*, especially its four protozoan species (*Plasmodium falciparum*, *P. malariae*, *P. ovale*, and *P. vivax*). Approximately, half of the world's population including infants, children below 5 years of age, and pregnant women are at risk of malarial infection. It is estimated that in the year 2018, there were 228 million cases of malaria globally as per the WHO Report 2019 on malaria (https://www.who.int/news-room/fact-sheets/ detail/malaria, accessed on May 16, 2020). Significant development has been made to fight malaria like the use of insecticide-treated mosquito nets and artemisinin-based combination therapy. However, the development of resistance towards the past and present antimalarial drugs spotlights the need to continue research to stay one step ahead (Lenzi et al. 2018; Foko et al. 2019). Fungi contain a variety of mycochemicals with antiplasmodial activity and have been screened to develop new antimalarial drugs (Ibrahim et al. 2018; Kadhila et al. 2018). Investigations revealed the presence of compounds, namely, cordypyridones A–B (31–32), in *Polycephalomyces nipponicus* which are atropisomers of each other. These have shown potent in vitro antimalarial activity with IC_{50s} of 0.066 and 0.037 µg/mL, respectively (Isaka et al. 2001). An insect pathogenic fungus, *O. unilateralis*, produces bioactive naphthoquinones with potent antimalarial activities (Kittakoop et al. 1999; Wongsa et al. 2005). A water-soluble polysaccharide with antimalarial activity has been purified from the ascocarps of a wood inhabiting ascomycete fungus, *Bulgaria inquinans*. It is composed of sugars, namely, mannose (27.2%), glucose (15.5%), and galactose (57.3%), and has molecular weight of 7.4 kDa. This polysaccharide possesses marked antimalarial activity and enhances artesunate activity in malaria-bearing mice. It has also been found to increase macrophage phagocytosis and proliferation of splenic lymphocyte in malaria-bearing mice and normal mice (Bi et al. 2011). The ethyl acetate extract of *G. lucidum* have been found to be rich in lanostanes with moderate in vitro antiplasmodial activities showing IC_{50s} from 6 to 20 µM (Adams et al. 2010). A terpenoid named sterostrein (33) obtained from the cultures of *Stereum ostrea* possesses considerable antimalarial activity against *P. falciparum* with IC_{50} of 2.3 µg/mL (Isaka et al. 2011, 2012). *Neonothopanus nambi* (*Marasmiaceae*) is a poisonous luminescent mushroom consisting of aristolane dimeric sesquiterpene and aurisin A which display antiparasitic properties. Aurisin A (34) and aurisin K (35) (Fig. 7.6; 31–35) terpenoids isolated from *N. nambi* possess excellent antimalarial activity against *Plasmodium falciparum* as well as antimycobacterial activity against *Mycobacterium tuberculosis* (Kanokmedhakul et al. 2012). Three nortriterpenes, named ganobonin-ketals (A–C), isolated from the basidiocarps of *G. orbiforme* display antiplasmodial activity with IC_{50} ranging from 1.7 to 7.9 µM (Ma et al. 2014a, b). Furthermore, ganodermalactones F, colossolactone E, and schisanlactone B identified in *Ganoderma* sp. KM01 culture exhibit modest antiplasmodial efficacy with IC_{50} in the range of 6.00–10.0 µM (Lakornwong et al. 2014). In Nigeria, mushrooms have

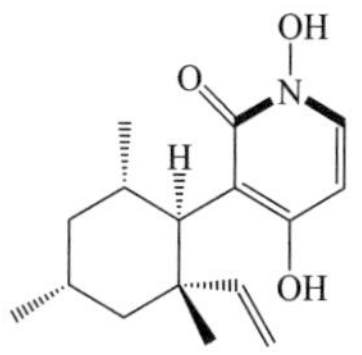

cordypyridone A

(31)

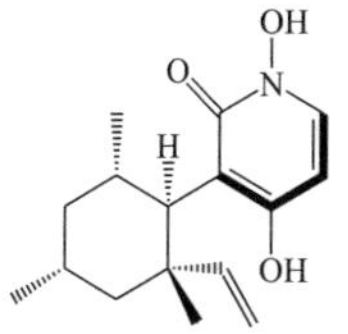

cordypyridone B

(32)

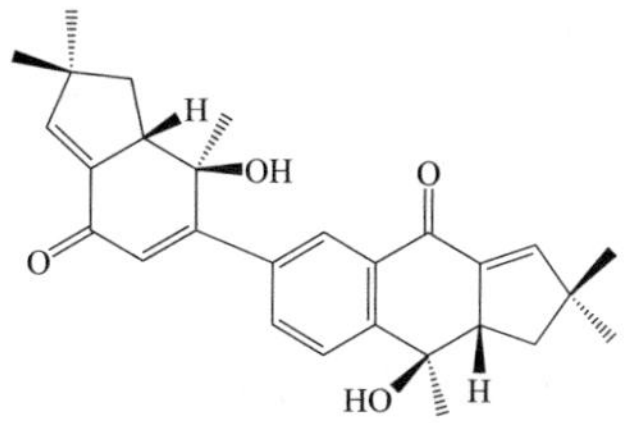

sterostrein A

(33)

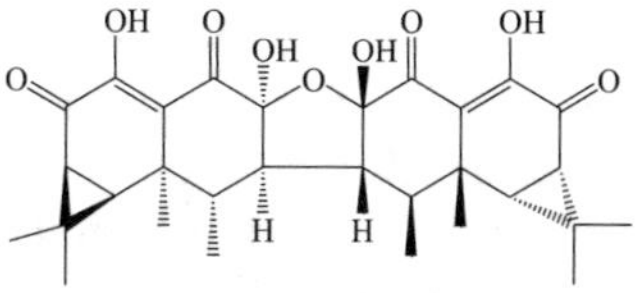

aurisin A

(34)

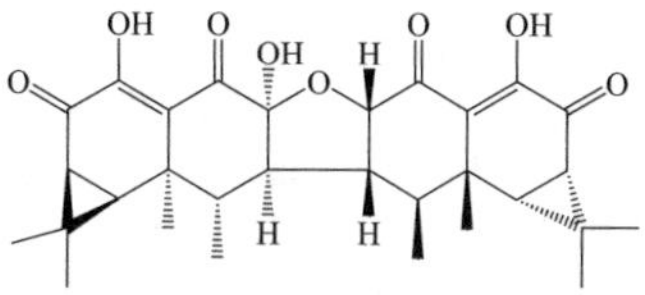

aurisin B

(35)

Fig. 7.6 Chemical structures of some antimalarial compounds from *P. nipponicus* (cordypyridone A and cordypyridone B), *Stereum ostrea* (sterostrein A), and *Neonothopanus nambi* (aurisin A and aurisin B)

been used to treat a number of ailments including the medicinal mushroom *Hypoxylon fuscum* which is generally found to be growing on dead tree trunks. The methanol extract of *H. fuscum* ascocarps contains secondary metabolites, including alkaloids, anthraquinones, cardiac glycosides, coumarins, flavonoids, saponins, sterols, and tannins. This extract has been observed with antiplasmodial activity against chloroquine-sensitive D6 and chloroquine-resistant W2 strains of *Plasmodium falciparum* with IC_{50s} of 6.98 and 8.33 µg/mL, respectively (Ogbole et al. 2018). Bioassay-guided compound isolation has resulted in the discovery of two new scalarane sesterterpenes and two new triterpenes from *P. ostreatus* and *S. areolatum* with antiprotozoan activity (Annang et al. 2018). The terpenoid extract generated from the basidiocarps of *G. lucidum* displays antiplasmodial action through its hypolipidemic activity. Its combination with chloroquine potentiates its curative effect in *Plasmodium berghei*-infected mice, lowering erythrocyte and hepatic lipids (Oluba 2019). Two new lanostane triterpenes, namely, ganoderic acid AW1 and ganoderic acid AW2, along with ganomycine A, pinellic acid, and ergosterol peroxide have been isolated from the basidiocarps of the cultivated new isolate of the Egyptian *Ganoderma* sp. Ganoderic acid AW1 shows significant antimalarial activity against *Plasmodium falciparum* (the chloroquine-sensitive strain) with IC_{50} of 257.8 nM without causing cytotoxicity up to the concentration of 9 µM (Wahba et al. 2019).

Antiviral Activity

Viral diseases are seriously dangerous for human health causing millions of people to suffer around the world. Presently, millions of people across the globe are suffering from the deadly coronavirus disease (COVID-19), and the WHO has declared it a pandemic (https://www.who.int/health-topics/coronavirus accessed May 25, 2020). To treat viral disorders, various antiviral drugs are in use against influenza virus (Principi et al. 2019), herpes simplex virus (HSV) (Tzeng et al. 2018), polio virus (PV) (McKinlay et al. 2014), human immunodeficiency virus (HIV) (Kenworthy et al. 2018), dengue virus (Botta et al. 2018), and coronavirus (Dong et al. 2020). Macrofungi represent a rich source of antiviral components effective against dengue, HSV, HIV, influenza, PV, and even the deadly coronaviruses (Elkhateeb et al. 2019a, b; Suwannarach et al. 2020). These antiviral mycochemicals obtained from various fungal taxa belonging to different orders of *Ascomycota* and *Basidiomycota* can be categorized into high (polysaccharides, polysaccharide-protein/amino acid complex, and lignin derivatives) and low molecular weight (secondary metabolites: alkaloids, polyketides, terpenoids, etc.) compounds (Linnakoski et al. 2018). For novel effective antiviral drugs, polysaccharides are effective and low toxic antiviral components (Chen and Huang 2018). The crude extracts and an isolated polysaccharide from the basidiocarps of *A. brasiliensis*, a native of Southeast Brazil, have shown antiviral activity against PV type 1 (PV-1) in HEp-2 cells. They are thought to inhibit the replication of this virus as they are effective at the time of infection (Faccin et al. 2007). Aqueous and ethanol extracts, as well as polysaccharide fractions of *L. edodes*, have potential against PV-1 and bovine herpes virus (BoHV-1) (Rincão et al. 2012). A novel lentinan (LNT-I) polysaccharide have been extracted from *L. edodes* mycelia. Structurally, it is made up of β-(1 → 3)-glucan backbone with β-(1 → 6)-glucosyl side-branching units terminating mannosyl and galactosyl residues and molecular weight of 3.79 × 105 Da. LNT-I consists of glucose, mannose, and galactose. LNT-I polysaccharide has been found with prominent antiviral activity against infectious hematopoietic necrosis virus (IHNV) causing direct inactivation and inhibition of viral replication. Additionally, it exerts immunomodulatory effects by downregulating the expression levels of TNF-α, IL-2, and IL-11 and up-modulating the expression levels of IFN-1 and IFN-γ after IHNV infection (Ren et al. 2018). Ribonucleases of fungal origin exhibit antiviral potential against human immunodeficiency virus type 1 (HIV-1) as reported with *Panellus serotinus* and *Corallium formosum* ribonucleases (Zhang et al. 2014a; Zhang et al. 2015). The ribosome-inactivating proteins have the capacity to inactivate ribosomes by the removal of one or more adenosine residues from rRNA and also inhibit HIV-1 reverse transcriptase activity and fungal proliferation (Puri et al.

2012). Laccases are related to pathogenesis, immune genesis, and morphogenesis of organisms (Xu et al. 2011). *P. cornucopiae* and *P. eryngii* laccases perform antiviral action on HIV-1 (Sun et al. 2014). The dichloromethane extract of *G. lucidum* rich in flavonoids, terpenoids, phenolics, and alkaloids have shown antihuman papilloma virus 16 (HPV 16) E6 onco-protein inhibition activity in epidermoid cervical carcinoma (CaSki) cells (Lai et al. 2010). A novel illudane-illudane bis-sesquiterpene, agrocybone (36), from *Agrocybe salicaceicola* exhibits antiviral activity against respiratory syncytial virus (RSV) with IC_{50} value of 100 μM (Zhu et al. 2010). Triterpenoids, namely, ganodermadiol (37), lucidadiol (38), and applanoxidic acid G (39) (Fig. 7.7; 36–39), isolated from *G. cupreolaccatum* display in vitro activity against influenza virus type A with IC_{50s} in Madin-Darby canine kidney cells (MDCK) > 0.22, 0.22, and 0.19 mM, respectively. Moreover, ganodermadiol acts against herpes simplex virus type 1 (HSV-1) causing lip exanthema and other symptoms with IC_{50} of 0.068 mM in Vero cells (Mothana et al. 2003). Many lanostane triterpenes, namely, colossolactones, have been extracted from the basidiocarps of *Ganoderma colossum*. These have antiviral activities against HIV-1 virus and suppress HIV-1 protease with IC_{50s} 5–39 μg/mL with colossolactone V and below 10 μg/mL with colossolactone G and schisanlactone A (El Dine et al. 2008). *Ganoderma sinense* triterpenoids, ganoderic acid GS-2, 20-hydroxylucidenic acid N, 20(21)-dehydrolucidenic acid N, and ganoderiol F exhibit antiviral effects

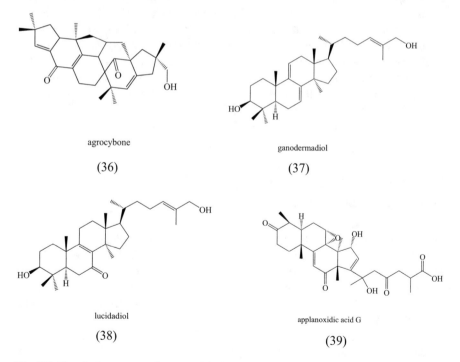

agrocybone

(36)

ganodermadiol

(37)

lucidadiol

(38)

applanoxidic acid G

(39)

Fig. 7.7 Chemical structures of some antiviral compounds from *Agrocybe salicacicola* (agrocybone) and *Ganoderma cupreolaccatum* (ganodermadiol, lucidadiol, and applanoxidic acid) against HIV-1 protease with IC_{50s} in the range of 20–40 μM (Sato et al. 2009). Earlier reports shed light on the relationship between oxidative stress and acquired immunodeficiency syndrome (AIDS), indicating the role of antioxidants in the treatment of AIDS (Foster 2007). Three antioxidant compounds, namely, adenosine, dimethylguanosine, and iso-sinensetin, from the ascocarps of *C. militaris* show moderate HIV-1 protease inhibiting potential (Jiang et al. 2011). It is well known that antiviral constituents act at different stages in the life cycle of the target virus and protect the host from infection. Aqueous extract of *A. blazei* acts directly on virus particles, namely, BoHV-1 and HSV-1 (Bruggemann et al. 2006), in the same way as an antiviral protein from *G. frondosa* acts on HSV (Gu et al. 2007) and aurenitol from *Chaetomium coarctatum* prevent adsorption of A (H3N2) (Sacramento et al. 2015) virus particles on the host cell surface. Another effective way to manage viral infections is the inhibition of viral replication. Polysaccharide-rich extracts of *Agaricus subrufescens* neither show virucidal action or prevent adsorption nor effect cells prior to infection. However, after infection, inhibit both virus strains, namely, HSV and bovine herpes virus, in HEp-2 cell cultures indicating their inhibitory effect on replication of these viruses (Yamamoto et al. 2013). Polysaccharopeptide complex from *T. versicolor* suppresses replication of HIV, influenza virus, and HSV (Collins and Ng, 1997; Krupodorova et al. 2014). Various other fungal bioactive metabolites target virus proteins, for example, velutin from *F. velutipes* targets HIV reverse transcriptase (Wang and Ng 2000) and 4.5 kDa protein of *Russula paludosa* inhibits HIV protease (Wang et al. 2007b). Ganoderic acid and triterpenoids of *G. lucidum* target enterovirus 71 and HIV (Min et al. 1998; Zhang et al. 2014b). Species in the genus *Ganoderma* possess various protease inhibitory compounds against HIV-1, e.g., ganomycin I and ganomycin B isolated from *Ganoderma colossus* (El Dine et al. 2009) and ganoderic acid B, ganoderic acid C1, ganoderic acid β, ganodermanondiol, ganodermanontriol, lucidumol B, and 3-5-dihydroxy-6-methoxyergosta-7,22-diene from *G. lucidum* (Tang et al. 2006b; Martinez-Montemayor et al. 2019).

Immunomodulatory, Neuroprotective, and Anti-inflammatory Activities

Several immunomodulatory, bioactive constituents of macrofungi participate in providing protection from cancer (Ayeka 2018), microbes (Mahamat et al. 2018), inflammation (Shao et al. 2019), etc. Polysaccharides from *G. lucidum* exhibit anti-Alzheimer's disease (Huang et al. 2017) and anti-inflammatory activities (Nagai et al. 2017), and those from *C. indica* and *F. velutipes* are neuroprotective (Govindan et al. 2014; Yang et al. 2015). *C. sinensis* and *Ganoderma atrum* polysaccharides produce immunomodulatory effects and provide in vivo protection from colon immune dysfunction induced by 150 mg/kg cyclophosphamide (CP) and also show anti-inflammatory properties. *C. sinensis* polysaccharides significantly increase the microbial-derived butyrate to improve histone h3 acetylation mediating regulatory Treg (T) cell-specific Foxp3 and restore CP-induced elevation of interleukin (IL)-17 and IL-21. Moreover, *G. atrum* polysaccharides exhibit significant downregulation of MyD88 and enhancement in IL-10 and TGF-β3. The mixture of fungal polysaccharides from *C. sinensis* and *G. atrum* balance the disequilibrium of cytokine secretion and Foxp3/RORγt ratio related Treg/T helper 17 (Th17) balance, downregulate the TLR-mediated inflammatory signaling pathway, and increase secretory immunoglobulin A (sIgA) secretion to inhibit colonic inflammation (Fan et al. 2018). Fungal immunomodulatory proteins found in *Stachybotrys chlorohalonata* and *F. velutipes* function as immunomodulatory as well as anti-inflammatory agents (Chu et al. 2017; Li et al. 2017a). Lectins (the nonimmune proteins or glycoproteins of the cell surface) in addition to antibacterial, antifungal, antitumor, and antiviral activities also exhibit various immunomodulatory properties (Singh et al. 2014). *P. ostreatus* and *A. bisporus* lectins produce immunomodulatory effects like activation of Toll-like receptor 6 signal pathway of dendritic cells and reduction of the innate and adaptive responses, respectively (Ditamo et al. 2016; He et al. 2017). Alzheimer's disease is a neurodegenerative disorder characterized by the accumulation of insoluble fibrillar senile plaques surrounding the neurons and deposition of hyperphosphorylated tau proteins (Duyckaerts et al. 2009). These plaques mainly consist of an amino acid peptide known as amyloid β-peptide (Aβ), which causes inadequacy of neurotransmitters, loss of neural functions, and death of neurons. All these events collectively lead to the development of Alzheimer's disease (Christen 2000; Smith et al. 2000; Sastre et al. 2006). Three labdane diterpenes derived from the basidiocarps of *T. camphoratus* show neuroprotective effects under in vitro studies (Chen et al. 2006). Some cyathane diterpenoids, namely, scabronines and sarcodonins, have been isolated from the basidiocarps of the basidiomycete *Sarcodon scabrosus*. Sarcodonins A (40) and G (41) at 25 µM concentration have been observed showing significant neurite outgrowth (neurite genesis)-promoting activities in the presence of 20 ng/mL NGF after 24 h treatment (Shi et al. 2011). It has been reported that scabronine M acts in a dose-dependent manner to inhibit NGF-induced neurite outgrowth in PC12 cells without causing any cytotoxic effect, probably by suppressing the phosphorylation of the receptor Trk A and the extracellular signal-regulated kinases (ERK) (Liu et al. 2012b). Further, cyathane diterpenes, namely, cyrneines A and B, from *Sarcodon cyrneus* induce neurite outgrowth in the PC12 cell model of neuronal differentiation, and cyrneine A even promotes neurite outgrowth in an Rac1-dependent mechanism in PC12 cells (Marcotullio et al. 2006; Obara et al. 2007). Four new benzofuran derivatives, ribisins A–D (42–45) (Fig. 7.8; 40–45), isolated from the methanolic extract of the basidiocarps of *Phellinus ribis* enhance neurite outgrowth in NGF-mediated PC12 cells 1 to 30 µM concentrations (Liu et al. 2012c). Terpenes isolated from *Cyathus striatus*, *C. africanus*, and *C. hookeri* also display neuroprotection (Han et al. 2013; Xu et al. 2013; Bai et al. 2015). Another major consequence of neurodegenerative diseases such as Alzheimer's, Parkinson's, Huntington's, and the prion diseases is neuronal cell death induced by endoplasmic reticulum (ER) stress. Hence, the compounds having the potential to low ER stress are helpful in the attenuation of neuronal cell death which in turn reduces damage in neurodegenerative disorders (Kawagishi et al. 1994; Shimoke et al. 2004). *H. erinaceus* is an edible and medicinal fungus

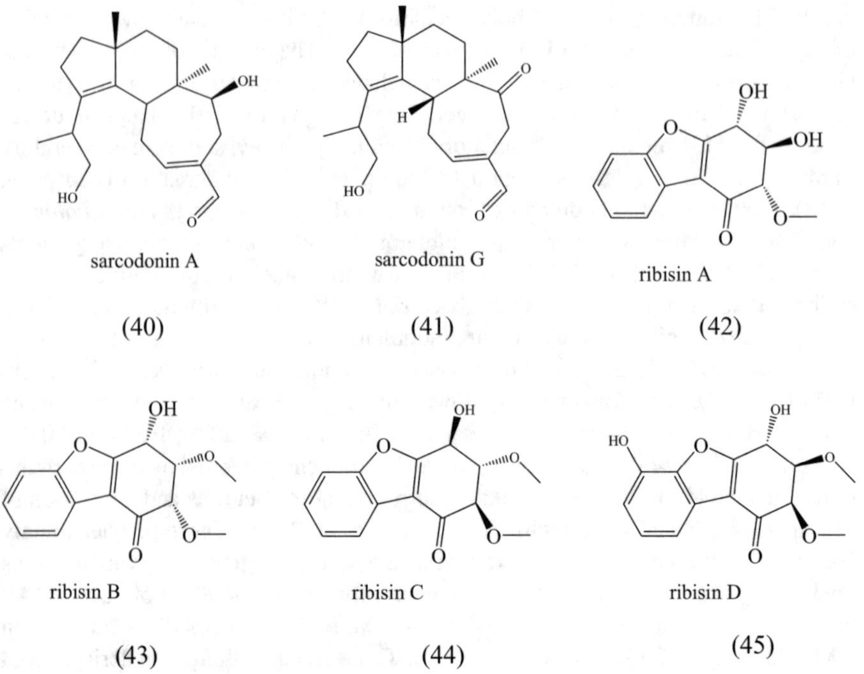

sarcodonin A

(40)

sarcodonin G

(41)

ribisin A

(42)

ribisin B

(43)

ribisin C

(44)

ribisin D

(45)

Fig. 7.8 Chemical structures of some bioactive compounds from *Sarcodon* species (sarcodonin A and sarcodonin G), benzofuran derivatives from *Phellinus ribis* (ribisin A, ribisin B, ribisin C, and ribisin D with neuroprotective effects

consisting of at least 2543 unique proteins, and by differential regulation of biosynthesis genes, this fungus could produce various bioactive metabolites with pharmacological effects (Zeng et al. 2018). Terpenoids, namely, hericenones and erinacines, isolated from *H. erinaceus* are able to cross the blood-brain barrier (Moldavan et al. 2007; Kawagishi and Zhuang 2008; Ma et al. 2010). Dilinoleoyl-phosphatidylethanolamine and 3-hydroxyhericenone F compounds isolated from *H. erinaceus* decrease ER stress-dependent neuronal cell death (Nagai et al. 2006; Ueda et al. 2008). In vivo studies conducted using erinacine A significantly enhances the level of nerve growth factor (NGF) in the rat's locus coeruleus and hippocampus but not in the cerebral cortex. This indicates medicinal characteristics of erinacines against degenerative nerve disorders and peripheral neuron regeneration (Shimbo et al. 2005). Phenolic compounds through their antioxidant potential protect the brain from oxidative stress produced under neurodegenerative disorders. Six phenolic compounds from *F. velutipes* have been observed to ameliorate H_2O_2-induced oxidative damage in PC12 cells and displayed neuroprotective potential along with their antioxidant action (Hu et al. 2017). The phenolic compounds present in *S. baumii*, an edible mushroom, have been observed to inhibit LPS-stimulated nitric oxide (NO) production in RAW 264.7 cells (Lee et al. 2017). Additionally, these phenolics also show neuroprotective potential due to their anti-acetylcholinesterase inhibitory activity as is reported for *G. lucidum* extracts (Cör et al. 2017). Linoleic acid, the most common polyunsaturated fatty acid in macrofungi, serves many physiological functions. It acts against inflammation by inhibiting the NO production and suppressing the expression of pro-inflammatory cytokines, TNF-α, IL-6, IL-1β, and NOS$_2$, in RAW 264.7 cells (Saiki et al. 2017) and protects from Alzheimer's disease risk through the inhibition of acetylcholinesterase and butyrylcholinesterase (Öztürk et al. 2014). Further, macrofungal species exhibit antimicrobial activities through the modulation of immune response. *T. clypeatus*, an edible mushroom growing commonly in different regions of Cameroon, shows bactericidal effect on five bacterial strains, viz., *Pasteurella multocida* NCTC 12178, methicillin-sensitive *S. aureus*, *E. coli*, *Enterobacter aerogenes*, and *Salmonella typhi*, and antifungal activities against two yeast species (*Candida albicans* and *C. glabrata*). The administration of *T. clypeatus* extract results in the delayed-type hypersensitivity response in healthy and dexamethasone immunosuppressed mice demonstrating the antibacterial and immunostimulatory activities of the extract on both cell-mediated and humoral immunity (Mahamat et al. 2018). Both edible and medicinal species of macrofungi have mycochemicals with excellent anti-inflammatory properties (Dua et al. 2018; Muszyńska et al. 2018). The ethanol extracts of six wild edible mushrooms, namely, *Craterellus cornucopioides*, *C. tubaeformis*, *Lactarius blennius*, *Russula fellea*, *R. mairei*, and hot water extract of *Pseudocraterellus undulatus* mushroom, exhibit anti-inflammatory potential in lipopolysaccharide-induced mouse RAW264.7 macrophage cell model of inflammation by lowering NO and IL-6 production (O'Callaghan et al. 2015). Nuclear factor-κB (NF-κB) plays important role in immune response and inflammation. Compounds isolated from *Ph. igniarius* (known as *Sang huang* in traditional Chinese medicine) cause inhibition of NF-κB in tumor necrosis factor-α-induced HeLa cells and protect from inflammation (Jiang et al. 2018). Total flavones recovered from the fermentation broth of the coculture of *C. comatus* and *M. esculenta* display anti-inflammtory effects in liposaccharide-challenged RAW264.7 macrophages via mitogen-activated protein kinase (MAPK) pathway. These flavone treatments lowers nitric oxide, TN-α and inteleukin-1β (IL-1β), and the activities of inducible NO synthase (iNOS) and cyclooxygenase 2 (COX-2) (Zhao et al. 2018b). Three acetylated polysaccharides (Ac-PMEP1-3) obtained from *Morchella angusticeps* exhibit immunomodulatory efficacy and anti-inflammatory action on RAW264.7 macrophages via NF-κB and p38/MAPK signaling pathways (Yang et al. 2019).

Miscellaneous Activities

Polysaccharides account for many health benefits, for example, those from *A. bisporus* display antiaging effect and protect from liver and nephric problems through the improvement of serum enzyme activities and modulating the levels of various biochemicals, lipids, and antioxidants (Li et al. 2017b). Antidiabetic, antioxidant, and anti-nephritic activities have been observed with *G. frondosa* polysaccharides.

G. frondosa polysaccharide mixtures (especially GFPS3 and GFPS4), have shown increase in body weight; lowering of plasma glucose; regulation of serum creatinine; blood urea nitrogen; N-acetyl-β-D-glucosaminidase and albuminuria; inhibition of interleukin (IL)-2, IL-6, and TNF-α levels in serum; improvement in the serum levels of matrix metalloproteinase 9 and interferon-α; and amelioration of pathological changes in the kidneys. Moreover, they have displayed modulations in the serum levels of SOD, GPx, CAT, MDA, and ROS as well as suppression of nuclear factor kappa B activities in the serum and kidneys (Kou et al. 2019). Bioactive fungal metabolites can be sustainably used to develop cosmeceutical or nutricosmetic formulations. Scientific data support that extracts made from medicinal, edible species and their individual metabolites possess antioxidant, anti-inflammatory, antimicrobial, photoprotective, anti-tyrosinase, anti-elastase, and anticollagenase activities. Therefore, these species can be employed as ingredients to get rid of hyperpigmentation, inflammatory skin disorders, and sun burning. However, stability, compatibility, and safety assessment and toxicological studies are still needed to be considered (Oludemi et al. 2020).

References

Adams M, Christen M, Plitzko I, Zimmermann S, Brun R, Kaiser M, Hamburger M (2010) Antiplasmodial lanostanes from the *Ganoderma lucidum* mushroom. J Nat Prod 73(5):897–900

Akihisa T, Uchiyama E, Kikuchi T, Tokuda H, Suzuki T, Kimura Y (2009) Anti-tumor-promoting effects of 25-methoxyporicoic acid A and other triterpene acids from *Poria cocos*. J Nat Prod 72(10):1786–1792

Alves M, Ferreira IFR, Dias J, Teixeira V, Martins A, Pintado M (2012) A review on antimicrobial activity of mushroom (*Basidiomycetes*) extracts and isolated compounds. Planta Medica 78(16):1707–1718

Alves MJ, Ferreira ICFR, Dias J, Teixeira V, Martins A, Pintado M (2013) A review on antifungal activity of mushroom (basidiomycetes) extracts and isolated compounds. Curr Top Med Chem 13(21):2648–2659

Ameri A, Vaidya JG, Deokule SS (2011) In vitro evaluation of antistaphylococcal activity of *Ganoderma lucidum*, *Ganoderma praelongum* and *Ganoderma resinaceum* from Pune, India. Afr J Microbiol Res 5(3):328–333

Annang F, Pérez-Victoria I, Appiah T, Pérez-Moreno G, Domingo E, Martín J, Mackenzie T, Ruiz-Pérez L, González-Pacanowska D, Genilloud O, Vicente F (2018) Antiprotozoan sesterterpenes and triterpenes isolated from two Ghanaian mushrooms. Fitoterapia 127:341–348

Antonio E, Alexander K, Florence L, Alessio C, Ramesh D, Marco E, Veronique M, Robert K (2015) Curr Med Chem 22(30):3502–3522

Arpha K, Phosri C, Suwannasai N, Mongkolthanaruk W, Sodngam S (2012) Astraodoric acids A-D: new lanostane triterpenes from edible mushroom *Astraeus odoratus* and their anti-*Mycobacterium tuberculosis* H37Ra and cytotoxic activity. J Agric Food Chem 60(39):9834–9841

Atta EM, Mohamed NH, Abdelgawad AAM (2017) Antioxidants: an overview on the natural and synthetic types. Eur Chem Bull 6(8):365–375

Ayala-Zavala JF, Silva-Espinoza BA, Cruz-Valenzuela MR, Villegas- Ochoa MA, Esqueda M, González-Aguilar GA, Calderón-López Y (2012) Antioxidant and antifungal potential of methanol extracts of *Phellinus* spp. from Sonora, Mexico. Rev Iberoam Micol 29(3):132–138

Ayeka PA (2018) Potential of mushroom compounds as immunomodulators in cancer immuno-therapy: a review. Evid-Based Complement Altern Med 2018:7271509

Azeem U (2017) Taxonomic studies on genus *Phellinus* from district Dehradun (Uttarakhand) and evaluation of some selected taxa for antihyperglycemic activity. PhD. Thesis, Department of Botany, Punjabi University, Punjab, India

Azeem U, Dhingra GS, Shri R (2018) Pharmacological potential of wood inhabiting fungi of genus *Phellinus* Quél.: an overview. J Pharmacogn Phytochem 7(2):1161–1171

Badalyan SM (2016) Fatty acid composition of different collections of coprinoid mush-rooms (Agaricomycetes) and their nutritional and medicinal values. Int J Med Mushrooms 18(10):883–893

Bai R, Zhang CC, Yin X, Wei J, Gao JM, Striatoids A–F (2015) Striatoids A–F, cyathane diter-penoids with neurotrophic activity from cultures of the fungus *Cyathus striatus*. J Nat Prod 78(4):783–788

Bal C (2019) Mushrooms as natural antimicrobial agents. Online J Complement Altern Med 2:3):1–3):3

Barreira JC, Oliveira MBP, Ferreira IC (2014) Development of a novel methodology for the analy-sis of ergosterol in mushrooms. Food Anal Method 7(1):217–223

Barrios-Garcia HB, Guizarnotegui-Blanco JA, Zapata-Campos CC, Almazan-Garcia C, Gonzalez-Alanis P, Villareal-Pena R, Hernandez-Jarguin A, Miranda-Hernandez DU, Martinez-Burnes J (2012) Identification of *Mycobacterium tuberculosis* complex by histopathology and PCR in white-tailed deer (*Odocoileus virginianus*) in Tamaulipas, Mexico. J Anim Vet Adv 11(7):1036–1040

Bhattarai G, Lee YH, Lee NH, Lee IK, Yun BS, Hwang PH, Yi HK (2012) Fomitoside-K from *Fomitopsis nigra* induces apoptosis of human oral squamous cell carcinomas (YD-10B) via mitochondrial signaling pathway. Biol Pharm Bull 35(10):1711–1719

Bi H, Han H, Li Z, Ni W, Chen Y, Zhu J, Gao T, Hao M, Zhou Y (2011) A water-soluble polysac-charide from the fruit bodies of *Bulgaria inquinans* (Fries) and its antimalarial activity. Evid-Based Complement Altern Med 2011:1–12

Blagodatski A, Yatsunskaya M, Mikhailova V, Tiasto V, Kagansky A, Katanaev VL (2018) Medicinal mushrooms as an attractive new source of natural compounds for future cancer therapy. Oncotarget 9(49):29259–29274

Botta L, Rivara M, Zuliani V, Radi M (2018) Drug repurposing approaches to fight Dengue virus infection and related diseases. Front Biosci Landmark 23:997–1019

Boucher HW, Corey GR (2008) Epidemiology of methicillin-resistant *Staphylococcus aureus*. Clin Infect Dis 46(5):S344–S349

Bruggemann R, Matsuo Orlandi J, Benati FJ, Faccin LC, Mantovani MS, Nozawa C, Linhares RE (2006) Antiviral activity of *Agaricus blazei* Murrill ss. Heinem extract against human and bovine herpesviruses in cell culture. Braz J Biol 37(4):561–565

Cao XY, Liu JL, Yang W, Hou X, Li QJ (2015) Antitumor activity of polysaccharide extracted from *Pleurotus ostreatus* mycelia against gastric cancer in vitro and in vivo. Mol Med Rep 12(2):2383–2389

Centko RM, Ramón-García R, Taylor T, Patrick BO, Thompson CJ, Miao VP, Andersen RJ (2012) Ramariolides A–D, antimycobacterial butenolides isolated from the mushroom *Ramaria cys-tidiophora*. J Nat Prod 75(12):2178–2182

Chandrasekaran G, Lee YC, Park H, Wu Y, Shin HJ (2016) Antibacterial and antifungal activi-ties of lectin extracted from fruiting bodies of the Korean cauliflower medicinal mushroom, *Sparassis latifolia* (Agaricomycetes). Int J Med Mushrooms 18(4):291–299

Chang ST, Wasser SP (2012) The role of culinary-medicinal mushrooms on human welfare with a pyramid model for human health. Int J Med Mushr 14(2):95–134

Chang HY, Sheu MJ, Yang CH, Lu TC, Chang YS, Peng WH, Huang SS, Huang GJ (2011) Analgesic effects and the mechanisms of anti-inflammation of hispolon in mice. Evid-Based Compl Altern Med 2011:1–9

Chaturvedi VK, Agarwal S, Gupta KK, Ramteke PW, Singh MP (2018) Medicinal mushroom: boon for therapeutic applications. 3 Biotech 8(8):334

Chen L, Huang G (2018) The antiviral activity of polysaccharides and their derivatives. Int J Biol Macromol 115:77–82

Chen CC, Shiao YJ, Lin RD, Shao YY, Lai MN, Lin CC, Ng LT, Kuo YH (2006) Neuroprotective diterpenes from the fruiting body of *Antrodia camphorata*. J Nat Prod 69(4):689–691

Chen H, Lu X, Qu Z, Wang Z, Zhang L (2010) Glycosidase inhibitory activity and antioxidant properties of a polysaccharide from mushroom *Inonotus obliquus*. J Food Biochem 34:178–191

Chen HP, Dong WB, Feng T, Yin X, Li ZH, Dong ZJ, Li Y, Liu JK (2014) Four new sesquiterpenoids from fruiting bodies of the fungus *Inonotus rickii*. J Asian Nat Prod Res 16(6):581–586

Chen L, Zhang Y, Sha O, Xu W, Wang S (2016) Hypolipidaemic and hypoglycaemic activities of polysaccharide from *Pleurotus eryngii* in Kunming mice. Int J Biol Macromol 93:1206–1209

Chen WY, Chang CY, Li JR, Wang JD, Wu CC, Kuan YH, Liao SL, Wang WY, Chen CJ (2018a) Anti-inflammatory and neuroprotective effects of fungal immunomodulatory protein involving microglial inhibition. Int J Mol Sci Nov 19(11):3678

Chen S, Su T, Wang Z (2018b) Structural characterization, antioxidant activity and immunological activity in vitro of polysaccharides from fruiting bodies of *Suillus granulatus*. J Food Biochem 43(2):1–9

Cheung PC (2013) Mini-review on edible mushrooms as source of dietary fiber: preparation and health benefits. Food Sci Hum Welln 2(3–4):162–166

Christen Y (2000) Oxidative stress and Alzheimer disease. Am J Clin Nutr 71(2):621S–629S

Chu PY, Sun HL, Ko JL, Ku MS, Lin LJ, Lee YT, Liao PF, Pan HH, Lu HL, Lue KH (2017) Oral fungal immunomodulatory protein-*Flammulina velutipes* has influence on pulmonary inflammatory process and potential treatment for allergic airway disease: a mouse model. J Microbiol Immunol Infect 50(3):297–306

Collins RA, Ng TB (1997) Polysaccharopeptide from *Coriolus versicolor* has potential for use against human immunodeficiency virus type 1 infection. Life Sci 60(25):383–387

Cör D, Botić T, Gregori A, Pohleven F, Knez Ž (2017) The effects of different solvents on bioactive metabolites and "in vitro" antioxidant and antiacetylcholinesterase activity of *Ganoderma lucidum* fruiting body and primordia extracts. Maced J Chem Chem Eng 36(1):129–141

Dan X, Liu W, Wong JH, Ng TB (2016) A ribonuclease isolated from wild *Ganoderma lucidum* suppressed autophagy and triggered apoptosis in colorectal cancer cells. Front Pharmacol 7:217

De Silva DD, Rapior S, Sudarman E, Stadler M, Jianchu X, Alias SA, Hyde KD (2013) Bioactive metabolites from macrofungi: ethnopharmacology, biological activities and chemistry. Fungal Divers 62(1):1–40

Dıaz-Godınez G (2015) Fungal bioactive compounds: an overview. Wiley Online Library

Diksha S, Singh VP, Singh NK (2018) A review on phytochemistry and pharmacology of medicinal as well as poisonous mushrooms. Mini Rev Med Chem 18(13):1095–1109

Ditamo Y, Rupil LL, Sendra VG, Nores GA, Roth GA, Irazoqui FJ (2016) In vivo immunomodulatory effect of the lectin from edible mushroom *Agaricus bisporus*. Food Funct 7(1):262–269

Dong L, Hu S, Gao J (2020) Discovering drugs to treat coronavirus disease 2019 (COVID-19). Drug Discov Ther 14(1):58–60

Du X, Zhang Y, Mu H, Lv Z, Yang Y, Zhang J (2015) Structural elucidation and antioxidant activity of a novel polysaccharide (TAPB1) from *Tremella aurantialba*. Food Hydrocoll 43:459–464

Dua B, Zhua F, Xu B (2018) An insight into the anti-inflammatory properties of edible and medicinal mushrooms. J Funct Foods 47:334–342

Dubey SK, Chaturvedi VK, Mishra D, Bajpeyee A, Tiwari A, Singh MP (2019) Role of edible mushroom as a potent therapeutics for the diabetes and obesity. 3 Biotech 9(12):450

Duru ME, Cayan GT (2015) Biologically active terpenoids from mushroom origin: a review. Rec Nat Prod 9(4):456–483

Duyckaerts C, Delatour B, Potier MC (2009) Classification and basic pathology of Alzheimer's disease. Acta Neuropathol 118(1):5–36

Dymkowska D, Drabarek B, Podszywałow-Bartnicka P, Szczepanowska J, Zabłocki K (2014) Hyperglycaemia modifies energy metabolism and reactive oxygen species formation in endothelial cells in vitro. Arch of Biochem Biophys 542:7–13

Ekowati N, Yuniati NI, Hernayanti RNI (2018) Antidiabetic potentials of button mushroom (*Agaricus bisporus*) on alloxan-induced diabetic rats. Biosaintifika 10(3):655–662

El Dine RS, El Halawany AM, Ma CM, Hattori M (2008) Anti-HIV-1 protease activity of lanostane triterpenes from the Vietnamese mushroom *Ganoderma colossum*. J Nat Prod 71(6):1022–1026

El Dine RS, El Halawany AM, Ma CM, Hattori M (2009) Inhibition of the dimerization and active site of HIV-1 protease by secondary metabolites from the Vietnamese Mushroom *Ganoderma colossum*. J Nat Prod 72(11):2019–2023

Elkhateeb WA, Daba GM, Elmahdy EM, Thomas PW, Wen T-C, Shaheen MNF (2019a) Antiviral potential of mushrooms in the light of their biological active compounds. ARC J Pharmac Sci 5(2):45–49

Elkhateeb WA, Daba GM, Thomas PW, Wen TC (2019b) Medicinal mushrooms as a new source of natural therapeutic bioactive compounds. Egypt Pharm J 18(2):88–101

Faccin LC, Benati F, Rincão VP, Mantovani MS, Soares SA, Gonzaga ML, Nozawa C, Carvalho Linhares RE (2007) Antiviral activity of aqueous and ethanol extracts and of an isolated polysaccharide from *Agaricus brasiliensis* against poliovirus type 1. Lett Appl Microbiol 45(1):24–28

Fan ST, Nie SP, Huang XJ, Wang S, Hu JL, Xie JH, Nie QX, Xie MY (2018) Protective properties of combined fungal polysaccharides from *Cordyceps sinensis* and *Ganoderma atrum* on colon immune dysfunction. Int J Biol Macromol 114:1049–1055

Fatmawati S, Shimizu K, Kondo R (2011) Ganoderol B: A potent α-glucosidase inhibitor isolated from the fruiting body of *Ganoderma lucidum*. Phytomed 18(12):1053–1055

Fernando D, Senathilake K, Nanayakkara C, de Silva ED, Wijesundera RL, Soysa P, De Silva N (2018) An insight in to isolation of natural products derived from macrofungi as antineoplastic agents: A Review. Int J Collab Res Int Med Public Health 10(1):816–828

Ferreira IC, Heleno SA, Reis FS, Stojkovic D, Queiroz MJ, Vasconcelos MH, Sokovic M (2015) Chemical features of *Ganoderma* polysaccharides with antioxidant, antitumor and antimicrobial activities. Phytochem 114:38–55

Flores AAU Jr, Alvarez MLC, Cortez FE, Perez BO, Sanico FL, Somoray MJM, Vicencio MCG, Cui KMR (2014) Inventory and utilization of macrofungi species for food and medicine. Int Conf Biol, Chem Environ Sci:25–28

Foko LPK, Meva FEA, Moukoko CEE, Ntoumba AA, Njila MIN, Kedi PBE, Ayong L, Lehman LG (2019) A systematic review on antimalarial drug discovery and antiplasmodial potential of green synthesis mediated metal nanoparticles: overview, challenges and future perspectives. Malaria J 18(1):337

Foster HD (2007) A role for the antioxidant defense system in preventing the transmission of HIV. Med Hypotheses 69(6):1277–1280

Gambato G, Pavão EM, Chilanti G, Fontana RC, Salvador M, Camassola M (2018) *Pleurotus albidus* modulates mitochondrial metabolism disrupted by hyperglycaemia in EA hy926 endothelial cells. BioMed Res Int 2018:1–11

Garcia K, Garcia CJ, Bustillos R, Dulay RM (2020) Mycelial biomass, antioxidant and mycoactives of mycelia of abalone mushroom *Pleurotus cystidiosus* in liquid culture. J Appl Biol Biotechnol 8(2):94–97

Gebreyohannes G, Nyerere A, Bii C, Sbhatu DB (2019) Determination of antimicrobial activity of extracts of indigenous wild mushrooms against pathogenic organisms. Evid-Based Complement Altern Med 2019:1–7

Giavasis I (2013) Production of microbial polysaccharides for use in food. In: Microbial production of food ingredients, enzymes and nutraceuticals. Woodhead Publishing, Cambridge, pp 413–468

Ginsberg AM (2010) Drugs in development for tuberculosis. Drugs 70(17):2201–2214

Govindan S, Keeper C, Jaba P, Rani JEE, Shanmugam J, Manoharan SKP (2014) *Calocybe indica* polysaccharides alleviate cognitive impairment, mitochondrial dysfunction and oxidative stress induced by D-galactose in mice. In: Proceedings of 8th international conference on mushroom biology and mushroom products, Volume I, II (2014) 394–406, ICAR-Directorate of Mushroom Research

Gu CQ, Li JW, Chao F, Jin M, Wang XW, Shen ZQ (2007) Isolation, identification and function of a novel anti-HSV-1 protein from *Grifola frondosa*. Antivir Res 75(3):250–257

Guillamón E, García-Lafuente A, Lozano M, Rostagno MA, Villares A, Martínez JA (2010) Edible mushrooms: role in the prevention of cardiovascular diseases. Fitoterapia 81(7):715–723

Gulati V, Singh MD, Gulati P (2019) Role of mushrooms in gestational diabetes mellitus. AIMS Med Sci 6(1):49–66

Halliwell B (2006) Reactive species and antioxidants. Redox biology is a fundamental theme of aerobic life. Plant Physiol 141(2):312–322

Han J, Chen Y, Bao L, Yang X, Liu D, Li S, Zhao F, Liu H (2013) Anti-inflammatory and cytotoxic cyathane diterpenoids from the medicinal fungus *Cyathus africanus*. Fitoterapia 8:422–431

Handa N, Yamada T, Tanaka R (2010) An unusual lanostane-type triterpenoid, spiroinonotsuoxodiol, and other triterpenoids from *Inonotus obliquus*. Phytochem 71(14–15):1774–1779

He M, Su D, Liu Q, Gao W, Kang Y (2017) Mushroom lectin overcomes hepatitis B virus tolerance via TLR6 signaling. Sci Rep 7(1):1–11

Heleno SA, Barros L, Martins A, Queiroz MJ, Santos-Buelga C, Ferreira IC (2012) Phenolic, polysaccharidic, and lipidic fractions of mushrooms from Northeastern Portugal: chemical compounds with antioxidant properties. J Agric Food Chem 60(18):4634–4640

Heleno SA, Stojković D, Barros L, Glamočlija J, Sokovićc M, Anabela M, QueirozbIsabe MJRP, Ferreiraa CFRI (2013) A comparative study of chemical composition, antioxidant and antimicrobial properties of *Morchella esculenta* (L.) Pers. from Portugal and Serbia. Food Research International 51(1):236–243

Heleno SA, Martins A, Queiroz MJR, Ferreira IC (2015a) Bioactivity of phenolic acids: metabolites versus parent compounds: a review. Food Chem 173:501–513

Heleno SA, Barros L, Martins A, Morales P, Fernández-Ruiz V, Glamoclija J, Sokovic M, Ferreira IC (2015b) Nutritional value, bioactive compounds, antimicrobial activity and bioaccessibility studies with wild edible mushrooms. LWT-Food Sci Technol 63(2):799–806

Hoeksma J, Misset T, Wever C, Kemmink J, Kruijtzer J, Versluis K, Liskamp RMJ, Boons GJ, Heck AJR, Boekhout T, Hertog JD (2019) A new perspective on fungal metabolites: identification of bioactive compounds from fungi using zebrafish embryogenesis as read-out. Sci Rep 9(1):1–16

Hsin IL, Ou CC, Wu MF, Jan MS, Hsiao YM, Lin CH, Ko JL (2015) GMI, an immunomodulatory protein from *Ganoderma microsporum*, potentiates cisplatin-induced apoptosis via autophagy in lung cancer cells. Mol Pharm 12(5):1534–1543

Hsin IL, Wang SC, Li JR, Ciou TC, Wu CH, Wu HM, Ko JL (2016) Immunomodulatory proteins FIP-gts and chloroquine induce caspase-independent cell death via autophagy for resensitizing cisplatin-resistant urothelial cancer cells. Phytomedicine 23(13):1566–1573

Hu Q, Wang D, Yu J (2017) Neuroprotective effects of six components from *Flammulina velutipes* on H2O2-induced oxidative damage in PC12 cells. J Funct Foods 37:586–593

Huang GJ, Yang CM, Chang YS, Amagaya S, Wang HC, Hou WC, Huang SS, Hu ML (2010) Hispolon suppresses SKHep1 human hepatoma cell metastasis by inhibiting matrix metalloproteinase-2/9 and urokinase-plasminogen activator through the PI3K/Akt and ERK signaling pathways. J Agric Food Chem 58(17):9468–9475

Huang GJ, Hsieh WT, Chang HY, Huang SS, Lin YC, Kuo YH (2011) α- Glucosidase and aldose reductase inhibitory activities from the fruiting body of *Phellinus merrillii*. J Agric Food Chem 59:5702–5706

Huang S, Mao J, Ding K, Zhou Y, Zeng X, Yang W, Wang P, Zhao C, Yao J, Xia P, Pei G (2017) Polysaccharides from *Ganoderma lucidum* promote cognitive function and neural progenitor proliferation in mouse model of Alzheimer's disease. Stem Cell Rep 8(1):84–94

Hyvönen ME, Dumont V, Tienari J, Lehtonen E, Ustinov J, Havana M, Jalanko H, Otonkoski T, Miettinen PJ, Lehtonen S (2015) Early-onset diabetic E1-DN mice develop albuminuria and glomerular injury typical of diabetic nephropathy. Biomed Res Int 2015:1–11

Ibrahim SRM, Mohamed GA, Al Haidari RA, El-Kholy AA, Zayed MF (2018) Potential anti-malarial agents from endophytic fungi: a review. Mini Rev Med Chem 18(13):1110–1132

Isaka M, Tanticharoen M, Kongsaeree P, Thebtaranonth Y (2001) Structures of cordypyridones A-D, antimalarial N-hydroxy- and Nmethoxy-2-pyridones from the insect pathogenic fungus *Cordyceps nipponica*. J Org Chem 66(14):4803–4808

Isaka M, Srisanoh U, ChoowongW BT (2011) Sterostreins A-E, new terpenoids from cultures of the basidiomycete *Stereum ostrea* BCC 22955. Org Lett 13(18):4886–4889

Isaka M, Srisanoh U, Sappan M, Supothina S, Boonpratuang T (2012) Sterostreins F-O, illu-dalanes and norilludalanes from cultures of the basidiomycete *Stereum ostrea* BCC 22955. Phytochem 79:116–120

Isaka M, Chinthanom P, Kongthong S, Srichomthong K, Choeyklin R (2013) Lanostane triter-penes from cultures of the basidiomycete *Ganoderma orbiforme* BCC 22324. Phytochemistry 87:133–139

Jang JS, Lee JS, Lee JH, Kwon DS, Lee KE, Lee SY, Hong EK (2010) Hispidin produced from *Phellinus linteus* protects pancreatic β-cells from damage by hydrogen peroxide. Arch Pharm Res 33:853–861

Jaworska G, Pogoń K, Bernaś E, Duda-Chodak A (2015) Nutraceuticals and antioxidant activity of prepared for consumption commercial mushrooms *Agaricus bisporus* and *Pleurotus ostreatus*. J Food Qual 38(2):111–122

Jiang Y, Wong JH, Fu M, Ng TB, Liu ZK, Wang CR, Li N, Qiao WT, Wen TY, Liu F (2011) Isolation of adenosine, isosinensetin and dimethylguanosine with antioxidant and HIV-1 protease inhibiting activities from fruiting bodies of *Cordyceps militaris*. Phytomed 18(2–3):189–193

Jiang Z, Jin M, Zhou W, Li R, Zhao Y, Jin X, Li G (2018) Anti-inflammatory activity of chemi-cal constituents isolated from the willow bracket medicinal mushroom *Phellinus igniarius* (Agaricomycetes). Intl J Med Mushrooms 20(2):119–128

Johansson M, Sterner O, Labischinski H, Anke T (2001) Coprinol, a new antibiotic cuparane from a *Coprinus* species. Z Naturforsch C 56(1–2):31–34

Jung JY, Lee IK, Seok SJ, Lee HJ, Kim YH, Yun BS (2008) Antioxidant polyphenols from the mycelial culture of the medicinal fungi Inonotus xeranticus and *Phellinus linteus*. J Appl Microbiol 104(6):1824–1832

Kalac P (2013) A review of chemical composition and nutritional value of wild-growing and cul-tivated mushrooms. J Sci Food Agric 93(2):209–218

Kalac P (2016) Edible mushrooms: chemical composition and nutritional value. Academic Press

Kang HS, Jun EM, Park SH, Heo SJ, Lee TS, Yoo ID, Kim JP (2007) Cyathusals A, B, and C, antioxidants from the fermented mushroom *Cyathus stercoreus*. J Nat Prod 70(6):1043–1045

Kanokmedhakul S, Lekphrom R, Kanokmedhakul K, Hahnvajanawong C, Bua-art S, Saksirirat W, Prabpai S, Kongsaeree P (2012) Cytotoxic sesquiterpenes from luminescent mushroom *Neonothopanus nimbi*. Tetrahedron 68:8261–8266

Kaur A, Dhingra GS, Shri R (2015) Antidiabetic potential of mushrooms. Asian J Pharm Res 5(2):111–125

Kawagishi H, Zhuang C (2008) Compounds for dementia from *Hericium erinaceum*. Drugs Fut 33(2):149

Kawagishi H, Shimada A, Shirai R, Okamoto K, Ojima F, Sakamoto H, Ishiguro Y, Furukawa S (1994) Hericenones A, B, C, strong stimulators of nerve growth factor synthesis, from the mycelia of *Hericium erinaceus*. Tetrahedr Lett 35:1569–1572

Kelner MJ, McMorris TC, Rojas RJ, Estes LA, Suthipinijtham P (2008) Synergy of irofulven in combination with other DNA damaging agents: synergistic interaction with altretamine, alkylating, and platinum-derived agents in the MV522 lung tumor model. Cancer Chemother Pharm 63:19–26

Kenworthy N, Thomann M, Parker R (2018) From a global crisis to the 'end of AIDS': New epidemics of signification. Glob Public Health 13(8):960–971

Khatua S, Paul S, Acharya K (2013) Mushroom as the potential source of new generation of antioxidant: a review. Res J Pharm Technol 6:496–505

Khatua S, Ghosh S, Acharya K (2017) Chemical composition and biological activities of methanol extract from *Macrocybe lobayensis*. J Appl Pharm Sci 7(10):144–151

Kim YJ, Park J, Min BS, Shim SH (2011) Chemical constituents from the sclerotia of *Inonotus obliquus*. J Korean Soc Appl Biol Chem 54(2):287–294

Kim KH, Moon E, Choi SU, Kim SY, Lee KR (2013) Lanostane triterpenoids from the mushroom *Naematoloma fasciculare*. J Nat Prod 76(5):845–851

Kinge TR, Mih AM (2011) Secondary metabolites of oil palm isolates of *Ganoderma zonatum* Murill. from Cameroon and their cytotoxicity against five human tumour cell lines. Afr J Biotechnol 10(42):8440–8447

Kittakoop P, Punya J, Kongsaeree P, Lertwerawat Y, Jintasirikul A, Tanticharoen M, Thebtaranonth Y (1999) Bioactive naphthoquinones from *Cordyceps unilateralis*. Phytochem 52:453–457

Kivrak I, Kivrak S, Harmandar M (2016) Bioactive compounds, chemical composition, and medicinal value of the giant puffball, *Calvatia gigantea* (higher basidiomycetes), from Turkey. Int J Med Mushrooms 18(2):97–107

Klaus A, Kozarski M, Vunduk J, Petrović P, Nikšić M (2017) Antibacterial and antifungal potential of wild basidiomycete mushroom *Ganoderma applanatum*. Lek Sirovine 36:37–46

Klein E, Smith DL, Laxminarayan R (2007) Hospitalizations and deaths caused by methicillin-resistant *Staphylococcus aureus*, United States, 1999–2005. Emerg Infect Dis 13(12):1840–1846

Kosanić M, Ranković B, Stanojković T, Radović-Jakovljević M, Ćirić A, Grujičić D, Milošević-Djordjević O (2019) *Craterellus cornucopioides* edible mushroom as source of biologically active compounds. Nat Product Commun 14:1934578–19843610

Kothari D, Patel S, Kim S-K (2018) Anticancer and other therapeutic relevance of mushroom polysaccharides: a holistic appraisal. Biomed Pharmacother 105:377–394

Kou L, Du M, Liu P, Zhang B, Zhang Y, Yang P, Shang M, Wang X (2019) Anti-diabetic and anti-nephritic activities of *Grifola frondosa* mycelium polysaccharides in diet-streptozotocin-induced diabetic rats Via modulation on oxidative stress. Appl Biochem Biotechnol 187(1):310–322

Krupodorova T, Rybalko S, Barshteyn V (2014) Antiviral activity of basidiomycete mycelia against influenza type A (serotype H1N1) and herpes simplex virus type 2 in cell culture. Virol Sin 29:284–290

Kumar K (2015) Role of edible mushrooms as functional foods – a review. South Asian J Food Technol Environ 1:211–218

Kumaran S, Pandurangan AK, Shenbhagaraman R, Esa NM (2017) Isolation and characterization of lectin from the artist's conk medicinal mushroom, *Ganoderma applanatum* (Agaricomycetes) and evaluation of its antiproliferative activity in HT-29 colon cancer cells. Int J Med Mushrooms 19(8):675–684

Lai LK, Abidin NZ, Abdullah N, Sabaratnam V (2010) Anti-human papillomavirus (HPV) 16E6 activity of Ling Zhi or Reishi medicinal mushroom, *Ganoderma lucidum* (W. Curt.: Fr.) P. Karst. (Aphyllophoromycetideae) extracts. Int J Med Mushrooms 12(3):279–286

Lakornwong W, Kanokmedhakul K, Kanokmedhakul S, Kongsaeree P, Prabpai S, Sibounnavong P, Soytong K (2014) Triterpene lactones from cultures of *Ganoderma* sp. KM01. J Nat Prod 77:1545–1553

Lee IK, Yun BS (2007) Highly oxygenated and unsaturated metabolites providing a diversity of hispidin class antioxidants in the medicinal mushrooms *Inonotus* and *Phellinus*. Bioorg Med Chem 15(10):3309–3314

Lee IK, Yun BS (2011) Styrylpyrone-class compounds from medicinal fungi *Phellinus* and *Inonotus* spp. and their medicinal importance. J Antibiot 64:349–359

Lee IK, Kim YS, Jang YW, Jung JY, Yun BS (2007a) New antioxidant polyphenols from the medicinal mushroom *Inonotus obliquus*. Bioorg Med Chem Lett 17(24):6678–6681

Lee IK, Kim YS, Seok SJ, Yun BS (2007b) Inoscavin E, a free radical scavenger from the fruiting bodies of *Inonotus xeranticus*. J Antibiot 60(12):745–747

Lee YS, Kang YH, Jung JY, Kang IJ, Han SN, Chung JS, Shin HK, Lim SS (2008) Inhibitory constituents of aldose reductase in the fruiting body of *Phellinus linteus*. Biol Pharm Bull 31(4):765–768

Lee YS, Kang IJ, Won MH, Lee JY, Kim JK, Lim SS (2010) Inhibition of protein tyrosine phosphatase 1beta by hispidin derivatives isolated from the fruiting body of *Phellinus linteus*. Nat Prod Commun 5(12):1927–1930

Lee IK, Jung JY, Yeom JH, Ki DW, Lee MS, Yeo WH, Yun BS (2012) Fomitoside K, a new lanostane triterpene glycoside from the fruiting body of *Fomitopsis nigra*. Mycobiology 40(1):76–78

Lee S, Lee D, Jang TS, Kang KS, Nam JW, Lee HJ, Kim KH (2017) Anti-inflammatory phenolic metabolites from the edible fungus *Phellinus baumii* in LPS-stimulated RAW 264.7 cells. Molecules 22(10):1583

Lenzi J, Costa TM, Alberton MD, Goulart JAG, Tavares LBB (2018) Medicinal fungi: a source of antiparasitic secondary metabolites. Appl Microbiol Biotechnol 102:5791–5810

Li H (2017) Extraction, purification, characterization and antioxidant activities of polysaccharides from *Ramaria botrytis* (Pers.) Ricken. Chem Cent J 11(1):24

Li YB, Liu RM, Zhong JJ (2013) A new ganoderic acid from *Ganoderma lucidum* mycelia and its stability. Fitoterapia 84:115–122

Li S, Jiang Z, Xu W, Xie Y, Zhao L, Tang X, Wang F, Xin F (2017a) FIP-sch2, a new fungal immunomodulatory protein from *Stachybotrys chlorohalonata*, suppresses proliferation and migration in lung cancer cells. Appl Microbiol Biotechnol 101(8):3227–3235

Li S, Liu H, Wang W, Wang X, Zhang C, Zhang J, Jing H, Ren Z, Gao Z, Song X, Jia L (2017b) Antioxidant and anti-aging effects of acidic-extractable polysaccharides by *Agaricus bisporus*. Int J Biol Macromol 106:1297–1306

Lin CH, Sheu GT, Lin YW, Yeh CS, Huang YH, Lai YC, Chang JG, Ko JL (2010) A new immunomodulatory protein from *Ganoderma microsporum* inhibits epidermal growth factor mediated migration and invasion in A549 lung cancer cells. Process Biochem 45(9):1537–1542

Lindequist U, Rausch R, Füssel A, Hanssen HP (2010) Higher fungi in traditional and modern medicine. Med Monatsschr Pharm 33(2):40–48

Ling C, Gangliang H (2018) Antitumor activity of polysaccharides: an overview. Curr Drug Targets 19(1):89–96(8)

Linnakoski R, Reshamwala D, Veteli P, Cortina-Escribano M, Vanhanen H, Varpu M (2018) Antiviral agents from fungi: diversity, mechanisms and potential applications. Front Microbiol 9:2325

Liu XT, Winkler AL, Schwan WR, Volk TJ, Rott M, Monte A (2010a) Antibacterial compounds from mushrooms II: lanostane triterpenoids and an ergostane steroid with activity against *Bacillus cereus* isolated from *Fomitopsis pinicola*. Planta Med 76(5):464–466

Liu XT, Winkler AL, Schwan WR, Volk TJ, Rott MA, Monte A (2010b) Antibacterial compounds from mushrooms I: a lanostane-type triterpene and prenylphenol derivatives from *Jahnoporus hirtus* and *Albatrellus flettii* and their activities against *Bacillus cereus* and *Enterococcus faecalis*. Planta Med 76(2):182–185

Liu J, Shimizu K, Tanaka A, ShinobuW OK, Nakamura T, Kondo R (2012a) Target proteins of ganoderic acid DM provides clues to various pharmacological mechanisms. Sci Rep 2:905

Liu L, Shi XW, Zong SC, Tang JJ, Gao JM (2012b) Scabronine M, a novel inhibitor of NGF-induced neurite outgrowth from PC12 cells from the fungus *Sarcodon scabrosus*. Bioorg Med Chem Lett 22(7):2401–2406

Liu Y, Kubo M, Fukuyama Y (2012c) Nerve growth factor-potentiating benzofuran derivatives from the medicinal fungus *Phellinus ribis*. J Nat Prod 75(12):2152–2157

Liu K, Xiao X, Wang J, Chen CY, Hu H (2017) Polyphenolic composition and antioxidant, antiproliferative and antimicrobial activities of mushroom *Inonotus sanghuang*. LWT-Food Sci Technol 82:154–161

Liu Y, Zhou Y, Liu M, Wang Q, Li Y (2018) Extraction optimization, characterization, antioxidant and immunomodulatory activities of a novel polysaccharide from the wild mushroom *Paxillus involutus*. Int J Biol Macromol 112:326–332

Liu Z, Zhao J-Y, Sun S-F, Li Y, Liu Y-B (2020) Fungi: outstanding source of novel chemical scaffolds. J Asian Nat Prod Res 22(2):99–120

Lu TL, Huang GJ, Lu TJ, Wu JB, Wu CH, Yang TC, Iizuka A, Chen YF (2009) Hispolon from *Phellinus linteus* has antiproliferative effects via MDM2-recruited ERK1/2 activity in breast and bladder cancer cells. Food Chem Toxicol 47:2013–2021

Ma BJ, Shen JW, Yu HY, Ruan Y, Wu TT, Zhao X (2010) Hericenones and erinacines: stimulators of nerve growth factor (NGF) biosynthesis in *Hericium erinaceus*. Mycol Int J Fungal Biol 1(2):92–98

Ma G, Yang W, Mariga AM, Fang Y, Ma N, Pei F, Hu Q (2014a) Purification, characterization and antitumor activity of polysaccharides from *Pleurotus eryngii* residue. Carbohydr Polym 114:297–305

Ma K, Ren J, Han J, Bao L, Li L, Yao Y, Sun C, Zhou B, Liu H (2014b) Ganoboninketals A–C, antiplasmodial 3,4-seco-27-norlanostane triterpenes from *Ganoderma boninense* Pat. J Nat Prod 77:1847–1852

Ma G, Yang W, Zhao L, Pei F, Fang D, Hu Q (2018) A critical review on the health promoting effects of mushrooms nutraceuticals. Food Sci Human Well 7:125–133

Mahamat O, André-Ledoux N, Chrisopher T, Mbifu AA, Albert K (2018) Assessment of antimicrobial and immunomodulatory activities of termite associated fungi, *Termitomyces clypeatus* R. Heim (Lyophyllaceae, Basidiomycota). Clin Phytosci 4(1):28

Marcotullio MC, Pagiotti R, Maltese F, Obara Y, Hoshino T, Nakahata N, Curini M (2006) Neurite outgrowth activity of cyathane diterpenes from *Sarcodon cyrneus*, cyrneines A and B. Planta Med 72(9):819–823

Martinez-Montemayor M, Ling T, Suárez-Arroyo IJ, Ortiz-Soto G, Santiago-Negrón CL, Lacourt-Ventura MY, Valentín-Acevedo A, Lang WH, Rivas F (2019) Identification of biologically active *Ganoderma lucidum* compounds and synthesis of improved derivatives that confer anticancer activities in vitro. Front Pharmacol 10:115

Matuszewska A, Karp M, Jaszek M, Janusz G, Osińska-Jaroszuk M, Sulej J, Stefaniuk D, Tomczak W, Giannopoulos K (2016) Laccase purified from *Cerrena unicolor* exerts antitumor activity against leukemic cells. Oncol Lett 11(3):2009–2018

Matuszewska A, Stefaniuk D, Jaszek M, Pięt M, Zając A, Matuszewski L, Cios I, Grąz M, paduch R, Bancerz R (2019) Antitumor potential of new low molecular weight antioxidative preparations from the white rot fungus *Cerrena unicolor* against human colon cancer cells. Sci Rep 9(1):1–10

McKinlay MA, Collett MS, Hincks JR, Steven Oberste M, Pallansch MA, Okayasu H, Sutter RW, Modlin JF, Dowdle WR (2014) progress in the development of poliovirus antiviral agents and their essential role in reducing risks that threaten eradication. J Infect Dis 210(1):S447–S453

McMorris TC (1999) Discovery and development of sesquiterpenoid derived hydroxymethylacylfulvene: a new anticancer drug. Bioorg Med Chem 7:881–886

McMorris TC, Kelner MJ, Wang W, Yu J, Estes LA, Taetle R (1996) (Hydroxymethyl) acylfulvene: an illudin derivative with superior antitumor properties. J Nat Prod 59:896–899

Min BS, Nakamura N, Miyashiro H, Bae KW, Hattori M (1998) Triterpenes from the spores of *Ganoderma lucidum* and their activity against HIV-1 protease. Chem Pharm Bull 46(10):1607–1612

Misiek M, Hoffmeister D (2012) Sesquiterpene aryl ester natural products in North American *Armillaria* species. Mycol Prog 11:7–15

Mizuno M, Nishitani Y (2013) Immunomodulating compounds in basidiomycetes. J Clin Biochem Nutr 52:202–207

Mo S, Wang S, Zhou G, Yang Y, Li Y, Chen X, Shi J (2004) Phelligridins C-F: cytotoxic pyrano[4,3-c][2]benzopyran-1,6-dione and furo[3,2-c]pyran-4-one derivatives from the fungus *Phellinus igniarius*. J Nat Prod 67:823–828

Moldavan MG, Gryganski AP, Kolotushkina OV, Kirchhoff B, Skibo GG, Pedarzani P (2007) Neurotropic and trophic action of lion's mane mushroom *Hericium erinaceus* (Bull.: Fr.) Pers. (Aphyllophoromycetideae) extracts on nerve cells in vitro. Int J Med Mushrooms 9(1):15–28

Mothana RAA, Awadh Ali NA, Jansen R, Wegner U, Mentel R, Lindequist U (2003) Antiviral lanostanoid triterpenes from the fungus *Ganoderma pfeifferi*. Fitoterapia 74(1):177–180

Muszyńska B, Grzywacz-Kisielewska A, Kała K, Gdula-Argasińska J (2018) Anti-inflammatory properties of edible mushrooms: a review. Food Chem 243:373–381

Mygind PH, Fischer RL, Schnorr KM, Hansen MT, Sönksen CP, Ludvigsen S, Raventós D, Buskov S, Christensen B, De Maria L, Taboureau O, Yaver D, Elvig-Jørgensen SG, Sørensen MV, Christensen BE, Kjærulff S, Frimodt-Moller N, Lehrer RI, Zasloff M, Kristensen HH (2005) Plectasin is a peptide antibiotic with therapeutic potential from a saprophytic fungus. Nature 437:975–980

Nagabushan H (2010) Retapamulin: a novel topical antibiotic. Indian J Dermatol Venereol Leprol (IJDVL) 76(1):77–79

Nagai K, Chiba A, Nishino T, Kubota T, Kawagishi H (2006) Dilinoleoyl-phosphatidylethanolamine from Hericium erinaceum protects against ER stress-dependent Neuro2a cell death via protein kinase C pathway. J Nutr Biochem 17:525–530

Nagai K, Ueno Y, Tanaka S, Hayashi R, Shinagawa K (2017) Polysaccharides derived from *Ganoderma lucidum* fungus mycelia ameliorate indomethacin-induced small intestinal injury via induction of GM-CSF from macrophages. Cell Immunol 320:20–28

Nakata T, Yamada T, Taji S, Ohishi H, Wada S, Tokuda H, Sakuma K, Tanaka R (2007) Structure determination of inonotsuoxides A and B and in vivo antitumor promoting activity of inotodiol from the sclerotia of *Inonotus obliquus*. Bioorg Med Chem 15:257–264

Nishikawa T, Edelstein D, Du XL, Yamagishi SI, Matsumura T, Kaneda Y, Yorek MA, Beebe D, Oates PJ, Hammes HP, Giardino I (2000) Normalizing mitochondrial superoxide production blocks three pathways of hyperglycaemic damage. Nature 404(6779):787–790

Nomura M, Takahashi T, Uesugi A, Tanaka R, Kobayashi S (2008) Inotodiol, a lanostane triterpenoid, from *Inonotus obliquus* inhibits cell proliferation through caspase-3-dependent apoptosis. Anticancer Res 28(5A):2691–2696

Novak R, Shlaes DM (2010) The pleuromutilin antibiotics: a new class for human use. Curr Opin Invest Dr 11(2):182–191

Obara Y, Hoshino T, Marcotullio MC, Pagiotti R, Nakahata N (2007) Novel cyathane diterpene, cyrneine A, induces neurite outgrowth in a Rac1-dependent mechanism in PC12 cells. Life Sci 80(18):1669–1677

Obodai M, Mensah DLN, Fernandes A, Kortei NK, Dzomeku M, Teegarden M, Schwartz SJ, Barros L, Prempeh J, Takli RK, Ferreira ICFR (2017) Chemical characterization and antioxidant potential of wild *Ganoderma* Species from Ghana. Molecules 22(196):1–18

O'Callaghan YC, O'Brien NM, Kenny O, Harrington T, Brunton N, Smyth TJ (2015) Anti-inflammatory effects of wild Irish mushroom extracts in RAW264. 7 mouse macrophage cells. J Med Food 18(2):202–207

Ogbole O, Segun P, Akinleye T, Fasinu P (2018) Antiprotozoal, antiviral and cytotoxic properties of the Nigerian mushroom, *Hypoxylon fuscum* Pers. Fr. (Xylariaceae). Acta Pharm Sci 56(4):2018

Oluba OM (2019) *Ganoderma* terpenoid extract exhibited anti-plasmodial activity by a mechanism involving reduction in erythrocyte and hepatic lipids in Plasmodium berghei infected mice. Lipids in Health Dis 18(12):1–9

Oludemi T, Filomena BM, Isabel CFRF (2020) The Role of bioactive compounds and other metabolites from mushrooms against skin disorders-a systematic review assessing their cosmeceutical and nutricosmetic outcomes. Curr Med Chem. Bentham Science Publishers

Othman L, Sleiman A, Abdel-Massih RM (2019) Antimicrobial activity of polyphenols and alkaloids in middle eastern plants. Front microbiol 10:911

Ou YX, Li YY, Qian XM, Shen YM (2012) Guanacastane-type diterpenoids from *Coprinus radians*. Phytochem 78:190–196

Ozen T, Kizil D, Yenigun S, Cesur H, Turkekul I (2019) Evaluation of bioactivities, phenolic and metal content of ten wild edible mushrooms from Western Black sea region of Turkey. Int J Med Mushrooms 21(10):979–994

Öztürk M, Tel G, Öztürk FA, Duru ME (2014) The cooking effect on two edible mushrooms in Anatolia: fatty acid composition, total bioactive compounds, antioxidant and anticholinesterase activities. Rec Nat Prod 8(2):189–194

Palacios I, Lozano M, Moro C, D'arrigo M, Rostagno MA, Martínez JA, García-Lafuente A, Guillamón E, Villares A (2011) Antioxidant properties of phenolic compounds occurring in edible mushrooms. Food Chem 128(3):674–678

Patel S, Goyal A (2012) Recent developments in mushrooms as anti-cancer therapeutics: a review. Biotechnol 2(1):1–15

Petrovska BB (2001) Protein fraction in edible Macedonian mushrooms. Eur Food Res Technol 212:469–472

Phan C-W, David P, Sabaratnam V (2018) Edible and medicinal mushrooms: emerging brain food for the mitigation of neurodegenerative diseases. J Med Food 20(1):1–10

Pires ADRA, Ruthes AC, Cadena SMSC, Iacomini M (2017) Cytotoxic effect of a mannogalacto-glucan extracted from *Agaricus bisporus* on HepG2 cells. Carbohydr Polym 170:33–42

Popova M, Trusheva B, Gyosheva M, Tsvetkova I, Bankova V (2009) Antibacterial triterpenes from the threatened wood-decay fungus *Fomitopsis rosea*. Fitoterapia 80(5):263–266

Prestinaci F, Pezzotti P, Pantosti A (2015) Antimicrobial resistance: a global multifaceted phenomenon. Pathog Glob Health 109(7):309–318

Principi N, Camilloni B, Alunno A, Polinori H, Argentiero A, Esposito S (2019) Drugs for influenza treatment: is there significant news? Front Med 6:109

Puri M, Kaur I, Perugini MA, Gupta RC (2012) Ribosome-inactivating proteins: current status and biomedical applications. Drug Discov 17(13–14):774–783

Rahi DK, Malik D (2016) Diversity of mushrooms and their metabolites of nutraceutical and therapeutic significance. J Mycol 2016:1–18

Rahvar A, Haas CS, Danneberg S, Harbeck B (2017) Increased cardiovascular risk in patients with adrenal insufficiency: a short review. Biomed Res Int 2017:1–6

Ramana KV, Reddy ABM, Majeti NVRK, Singhal SS (2018) Therapeutic potential of natural antioxidants. Oxi Med Cell Longev 2018:9471051

Rao YK, Wu AT, Geethangili M, Huang MT, Chao WJ, Wu CH, Deng WP, Yeh CT, Tzeng YM (2011) Identification of antrocin from *Antrodia camphorata* as a selective and novel class of small molecule inhibitor of Akt/mTOR signaling in metastatic breast cancer MDA-MB-231 cells. Chem Res Toxicol 24(2):238–245

Raseta M, Popovi M, Capo I, Stilinovi N, Vukmirovi S, Milosevi B, Karamand M (2020) Antidiabetic effect of two different *Ganoderma* species tested in alloxan diabetic rats. RSC Adv 10(17):10382–10393

Rathore H, Prasad S, Sharma S (2017) Mushroom nutraceuticals for improved nutrition and better human health: a review. Pharma Nutr 5(2):35–46

Ren D, Wang N, Guo J, Yuan L, Yang X (2016) Chemical characterization of *Pleurotus eryngii* polysaccharide and its tumor-inhibitory effects against human hepatoblastoma HepG-2 cells. Carbohydr Polym 138:123–133

Ren G, Xu L, Lu T, Yin J (2018) Structural characterization and antiviral activity of lentinan from Lentinus edodes mycelia against infectious hematopoietic necrosis virus. Int J Biol Mcromol 115:1202–1210

Rincão VP, Yamamoto KA, Ricardo NM, Soares SA, Meirelles LD, Nozawa C, Linhares REC (2012) Polysaccharides and extracts from *Lentinula edodes*: structural features and antiviral activity. Virol J 15:37

Ruthes AC, Smiderle FR, Iacomini M (2016) Mushroom heteropolysaccharides: a review on their sources, structure and biological effects. Carbohydr Polym 136:358–375

Sacramento CQ, Marttorelli A, Fintelman-Rodrigues N, de Freitas CS, de Melo GR, Rocha ME, Kaiser CR, Rodrigues KF, da Costa GL, Alves CM, Santos-Filho O (2015) Aureonitol, a fungi derived tetrahydrofuran, inhibits influenza replication by targeting its surface glycoprotein hemagglutinin. PLoS One 10:0139236

Saiki P, Kawano Y, Van Griensven LJ, Miyazaki K (2017) The anti-inflammatory effect of *Agaricus brasiliensis* is partly due to its linoleic acid content. Food Funct 8(11):4150–4158

Sanico FL, Somoray MJM, Alvarez MLC, Cortez FE, Perez BO, Flores, Jr AAU, Vicencio MCG, Cui KMR (2014) Determination of nutrients and nutraceutical content of wild mushroom species. International Conference on Emerging Trends in Computer and Image Processing (ICETCIP'2014) Dec. 15–16, 2014, Pattaya (Thailand)

Sastre M, Klockgether T, Heneka MT (2006) Contribution of inflammatory processes to Alzheimer's disease: molecular mechanisms. Int J Dev Neurosci 24(2–3):167–176

Sato N, Zhang Q, Ma C-M, Hattori M (2009) Anti-human immunodeficiency virus-1 protease activity of new lanostane- type triterpenoids from *Ganoderma sinense*. Chem Pharm Bull 57:1076–1080

Schobert R, Knauer S, Seibt S, Biersack B (2011) Anticancer active illudins: recent developments of a potent alkylating compound class. Curr Med Chem 18(6):790–807

Schüffler A (2018) Secondary metabolites of basidiomycetes. In: Physiology and genetics. The Mycota (A comprehensive treatise on fungi as experimental systems for basic and applied research), vol 15. Springer, Cham

Schüffler A, Wollinsky B, Anke T, Liermann JC, Opatz T (2012) Isolactarane and sterpurane sesquiterpenoids from the basidiomycete *Phlebia uda*. J Nat Prod 75(7):1405–1408

Sezgin S, Dalar A, Uzun Y (2020) Determination of antioxidant activities and chemical composition of sequential fractions of five. edible mushrooms from Turkey. J Food Sci Technol 57(5):1866–1876

Shao HJ, Jeong JB, Kim KJ, Lee SH (2015) Anti-inflammatory activity of mushroom derived hispidin through blocking of NF-$_k$B activation. J Sci Food Agric 95(!2):2482–2486

Shao KD, Mao PW, Li QZ, Li LDJ, Wang YL, Zhou XW (2019) Characterization of a novel fungal immunomodulatory protein, FIP-SJ75 shuffled from *Ganoderma lucidum*, *Flammulina velutipes* and *Volvariella volvacea*. Food Agric Immunol 30(1):1253–1270

Sharma SK, Gautam N, Atri NS (2015) Optimized extraction, composition, antioxidant and antimicrobial activities of exo and intracellular polysaccharides from submerged culture of *Cordyceps cicadae*. BMC Complement Altern Med 15:446

Shen M, Zhiyu F, Yutao C, Yidan C, Bin X, Li G, Yaoyao X, Ge W, Weimin W, Yongjun Z (2019) Hypoglycemic effect of the degraded polysaccharides from the wood ear medicinal mushroom *Auricularia auricula-judae* (Agaricomycetes). Int J Med Mushrooms 21(10):1033–1042

Shi XW, Liu L, Gao JM, Zhang AL (2011) Cyathane diterpenes from Chinese mushroom *Sarcodon scabrosus* and their neurite outgrowth promoting activity. Eur J Med Chem 46(7):3112–3117

Shimbo M, Kawagishi H, Yokogosh H (2005) Erinacine A increases catecholamine and nerve growth factor content in the central nervous system of rats. Nutr Res 25(6):617–623

Shimoke K, Amano H, Kishi S, Uchida H, Kudo M, Ikeuchi T (2004) Nerve growth factor attenuates endoplasmic reticulum stress-mediated apoptosis via suppression of caspase-12 activity. J Biochem 135(3):439–446

Sinanoglou VJ, Zoumpoulakis P, Heropoulos G, Proestos C, Ćirić A, Petrovic J, Glamoclija J, Sokovic M (2014) Lipid and fatty acid profile of the edible fungus *Laetiporus sulphureus*. Antifungal and antibacterial properties. J Food Sci Technol 52(6):3264–3272

Singh SS, Wang H, Chan YS, Pan W, Dan X, Yin CM, Akkouh O, Ng TB (2014) Lectins from edible mushrooms. Molecules 20(1):446–469

Singh V, Bedi GK, Shri R (2017) In vitro and in vivo antidiabetic evaluation of selected culinary-medicinal mushrooms (Agaricomycetes). Int J Med Mushrooms 19(1):17–25

Smith MA, Rottkamp CA, Nunomura A, Raina AK, Perry G (2000) Oxidative stress in Alzheimer's disease. Biochimica et Biophysica Acta (BBA)-Molecular Basis of Disease 1502(1):139–144

Srivastava S, Verma NK, Vishwakarma DK, Mishra JN (2018) An overview on mushroom: Chemical constituents and pharmacological activities. Int J Pharm Sci Res 4(1):56–65

Stanikunaite R, Radwan MM, Trappe JM, Fronczek F, Ross SA (2008) Lanostane-type triterpenes from the mushroom *Astraeus pteridis* with antituberculosis activity. J Nat Prod 71(12):2077–2079

Sun J, Chen QJ, Zhu MJ, Wang HX, Zhang GQ (2014) An extracellular laccase with antiproliferative activity from the sanghuang mushroom *Inonotus baumii*. J Mol Catal B: Enzym 99:20–25

Suseem SR, Saral AM (2013) Analysis on essential fatty acid esters of mushrooms *Pleurotus eous* and its antibacterial activity. Asian J Pharmaceut Clin Res 6(1):188–191

Suwannarach N, Kumla J, Sujarit K, Pattananandecha T, Saenjum C, Lumyong S (2020) Natural bioactive compounds from fungi as potential candidates for protease inhibitors and immunomodulators to apply for coronaviruses. Molecules 25(1800):1–21

Taji S, Yamada T, In Y, Wada S, Usami Y, Sakuma K, Tanaka R (2007) Three new lanostane triterpenoids from *Inonotus obliquus*. Helv Chim Acta 90(11):2047–2057

Taji S, Yamada T, Tanaka R (2008a) Three new lanostane triterpenoids, inonotsutriols A, B, and C from *Inonotus obliquus*. Helv Chim Acta 91(8):1513–1524

Taji S, Yamada T, Wada S, Tokuda H, Sakuma K, Tanaka R (2008b) Lanostane-type triterpenoids from the sclerotia of *Inonotus obliquus* possessing anti-tumor promoting activity. Eur J Med Chem 43(11):2373–2379

Tan X, Sun J, Xu Z, Li H, Hu J, Ning H, Qin Z, Pei H, Sun T, Zhang X (2018) Effect of heat stress on production and in vitro antioxidant activity of polysaccharides in *Ganoderma lucidum*. Bioproc Biosyst Eng 41(1):35–141

Tanaka R, Toyoshima M, Yamada T (2011) New lanostane triterpenoids, inonotsutriols D and E from *Inonotus obliquus*. Phytochem Lett 4(3):329–332

Tang W, Gu T, Zhong JJ (2006a) Separation of targeted ganoderic acids from *Ganoderma lucidum* by reversed phase liquid chromatography with ultraviolet and mass spectrometry detections. Biochem Eng J 32:205–210

Tang W, Liu JW, ZhaoWM WDZ, Zhong JJ (2006b) Ganoderic acid T from *Ganoderma lucidum* mycelia induces mitochondria mediated apoptosis in lung cancer cells. Life Sci 80:205–211

Tel-Cayan G, Ozturk M, Duru ME, Turkoglu A (2017) Fatty acid profiles in wild mushroom species from Anatolia. Chem Nat Compd 53(2):351–353

Teoh HL, Ahmad IS, Johari NMK, Aminudin N, Abdullah N (2019) Antioxidant properties and yield of wood ear mushroom, *Auricularia polytricha* (Agaricomycetes), cultivated on rubber-wood sawdust. Int J Med Mushrooms 20(4):369–380

Thu ZM, Myo KK, Aung HT, Clericuzio M, Armijos C, Vidari G (2020) Bioactive phytochemical constituents of wild edible mushrooms from Southeast Asia. Molecules 25(8):1972

Toledo CV, Barroetaveña C, Fernandes A, Barros L, Ferreira ICFR (2016) Chemical and antioxidant properties of wild edible mushrooms from native *Nothofagus* spp. Forest, Argentina. Molecules 21(1201):1–15

Kadhila NP, Sekhoacha M, Tselanyane M, Malefa L, Chinsembu KC (2018) Anti-plasmodial activities in mushrooms. Int J Vector Borne Dis 2018:139–142

Tu SH, Wu CH, Chen LC, Huang CS, Chang HW, Chang CH, Lien HM, Ho YS (2012) In vivo antitumor effects of 4,7-dimethoxy-5-methyl- 1,3-benzodioxole isolated from the fruiting body of *Antrodia camphorata* through activation of the p53-mediated p27/Kip1 signaling pathway. J Agric Food Chem 60(14):3612–3618

Tzeng MA, Chung CH, Lin FH, Chiang CP, Yeh CB, Huang SY, Lu RB, Chang HA, Kao YC, Yeh HW, Chiang WS, Chou YC, Tsao CH, Wu YF, Chien WC (2018) Anti-herpetic medications and reduced risk of dementia in patients with herpes simplex virus infections-a nationwide, population-based cohort study in Taiwan. Neurother 15(2):417–429

Ueda K, Tsujimori M, Kodani S, Chiba A, Kubo M, Masuno K, Sekiya A, Nagai K, Kawagishi H (2008) An endoplasmic reticulum (ER) stress-suppressive compound and its analogues from the mushroom *Hericium erinaceum*. Bioorg Med Chem 16(21):9467–9470

Vacca P, Fazio C, Neri A, Ambrosio L, Palmieri A, Stefanelli P (2018) Neisseria meningitidis antimicrobial resistance in Italy, 2006 to 2016. Antimicrob Agents Chemother 62(9)

Vasdekis EP, Karkabounas A, Giannakopoulos I, Savvas D, Lekka ME (2018) Screening of mushrooms bioactivity: piceatannol was identified as a bioactive ingredient in the order Cantharellales. Eur Food Res Technol 244(5):861–871

Wahba AE, El-Sayed AKA, El-Falal AA, Soliman EM (2019) New antimalarial lanostane triterpenes from a new isolate of Egyptian *Ganoderma* species. Med Chem Res 28(12):2246–2251

Wang H, Ng TB (2000) Isolation of a novel ubiquitin-like protein from *Pleurotus ostreatus* mushroom with anti-human immunodeficiency virus, translation-inhibitory and ribonuclease activities. Biochem Biophys Res Commun 276(2):587–593

Wang Y, Mo SY, Wang SJ, Li S, Yang YC, Shi JG (2005) A unique highly oxygenated Pyrano[4,3-c][2]benzopyran-1,6-dione derivative with antioxidant and cytotoxic activities from the fungus *Phellinus igniarius*. Org Lett 7(9):1675–1678

Wang Y, Shang XY, Wang SJ, Mo SY, Li S, Yang YC, Ye F, Shi JG, Lan H (2007a) Structures, biogenesis, and biological activities of Pyrano[4,3-c]isochromen-4-one derivatives from the fungus *Phellinus igniarius*. J Nat Prod 70(2):296–299

Wang J, Wang HX, Ng TB (2007b) A peptide with HIV-1 reverse transcriptase inhibitory activity from the medicinal mushroom *Russula paludosa*. Peptides 28(3):560–565

Wang Y, Bao L, Yang X, Li L, Li S, Gao H, Yao XS, Wen H, Liu HW (2012a) Bioactive sesquiterpenoids from the solid culture of the edible mushroom *Flammulina velutipes* growing on cooked rice. Food Chem 132(3):1346–1353

Wang SJ, Li YX, Bao L, Han JJ, Yang XL, Li HR, Wang YQ, Li SJ, Liu HW, Eryngiolide A (2012b) Eryngiolide A, A cytotoxic macrocyclic diterpenoid with an unusual cyclododecane core skeleton produced by the edible mushroom *Pleurotus eryngii*. Org Lett 14(14):3672–3675

Wang Y, Liu Y, Hu Y (2014) Optimization of polysaccharides extraction from *Trametes robiniophila* and its antioxidant activities. Carbohydr Polym 111(13):324–332

Wang J, Li W, Huang X, Liu Y, Li Q, Zheng Z, Wang KA (2017) A polysaccharide from *Lentinus edodes* inhibits human colon cancer cell proliferation and suppresses tumor growth in athymic nude mice. Oncotarget 8(1):610–623

Wasser SP (2011) Current findings, future trends, and unsolved problems in studies of medicinal mushrooms. Appl Microbiol Biotechnol 89(5):1323–1332

Wasser SP (2014) Medicinal mushroom science: current perspectives, advances, evidences and challenges. Biomed J 37(6):345–356

Wasser SP (2017) Medicinal mushrooms in human clinical studies. Part I. anticancer, oncoimmunological, and immunomodulatory activities: a review. Int J Med Mushrooms 19(4):279–317

Watkins RR, Holubar M, David MZ (2019) Antimicrobial resistance in methicillin-resistant *Staphylococcus aureus* to newer antimicrobial agents. Antimicrob Agents Chemother 63(12):e01216–e01219

Weng CJ, Chau CF, Chen KD, Chen DH, Yen GC (2007) The antiinvasive effect of lucidenic acids isolated from a new *Ganoderma lucidum* strain. Mol Nutr Food Res 51(12):1472–1477

Wongsa P, Tasanatai K, Watts P, Hywel-Jones N (2005) Isolation and in vitro cultivation of the insect pathogenic fungus *Cordyceps unilateralis*. Mycol Res 109(8):936–940

Woynarowski J, Napier C, Koester S (1997) Effects on DNA integrity and apoptosis induction by a novel antitumor sesquiterpene drug, 6-hydroxymethylacylfulvene (HMAF). Biochem Pharmacol 54(11):1181–1193

Wu GS, Lu JJ, Guo JJ, Li YB, Tan W, Dang YY, Zhong ZF, Xu ZT, Chen XP, Wang YT (2012) Ganoderic acid DM, a natural triterpenoid, induces DNA damage, G1 cell cycle arrest and apoptosis in human breast cancer cells. Fitoterapia 83(2):408–414

Xiao C, Wu Q, Zhang J, Xie Y, Cai W, Tan J (2017) Antidiabetic activity of *Ganoderma lucidum* polysaccharides F31 down-regulated hepatic glucose regulatory enzymes in diabetic mice. J Ethnopharmacol 196:47–57

Xu X, Yan H, Chen J, Zhang X (2011) Bioactive proteins from mushrooms. Biotechnol Adv 29(6):667–674

Xu Z, Yan S, Bi K, Han J, Chen Y, Wu Z, Liu H (2013) Isolation and identification of a new anti-inflammatory cyathane diterpenoid from the medicinal fungus *Cyathus hookeri* Berk. Fitoterapia 86:159–162

Xu S, Dou Y, Ye B, Wu Q, Wang Y, Hu M, Ma F, Rong X, Guo J (2017) *Ganoderma lucidum* polysaccharides improve insulin sensitivity by regulating inflammatory cytokines and gut microbiota composition in mice. J Funct Foods 38:545–552

Yahia EM, Gutiérrez-Orozco F, Moreno-Pérez MA (2017) Identification of phenolic compounds by liquid chromatography-mass spectrometry in seventeen species of wild mushrooms in Central Mexico and determination of their antioxidant activity and bioactive compounds. Food Chem 226:14–22

Yamamoto KA, Galhardi LC, Rincão VP, de Aguiar Soares S, Ricardo NM, Nozawa C, Linhares RE (2013) Antiherpetic activity of an *Agaricus brasiliensis* polysaccharide, its sulfated derivative and fractions. Int J Biol Macromol 52:9–13

Yang W, Yu J, Zhao L, Ma N, Fang Y, Pei F, Mariga AM, Hu Q (2015) Polysaccharides from *Flammulina velutipes* improve scopolamine-induced impairment of learning and memory of rats. J Funct Foods 18:411–422

Yang G, Yang L, Zhuang Y, Qian X, Shen Y (2016) *Ganoderma lucidum* polysaccharide exerts anti-tumor activity via MAPK pathways in HL-60 acute leukemia cells. J Recept Signal Transduct 36(1):6–13

Yang Y, Chen J, Lei L, Li F, Tang Y, Yuan Y, Zhang Y, Wu S, Yin R, Ming J (2019) Acetylation of polysaccharide from *Morchella angusticeps* peck enhances its immune activation and anti-inflammatory activities in macrophage RAW264. 7 cells. Food Chem Toxicol 125:38–45

Yeh CT, Rao YK, Yao CJ, Yeh CF, Li CH, Chuang SE, Luong JH, Lai GM, Tzeng YM (2009) Cytotoxic triterpenes from *Antrodia camphorata* and their mode of action in HT-29 human colon cancer cells. Cancer Lett 285(1):73–79

Yu Y, Shen M, Song Q, Xie J (2018) Biological activities and pharmaceutical applications of polysaccharide from natural resources: a review. Carbohydr Polym 183:91–101

Yuan Q, Zhao L, Li Z, Harqin C, Pengm Y, Liu J (2018) Physicochemical analysis, structural elucidation and bioactivities of a high-molecular-weight polysaccharide from *Phellinus igniarius* mycelia. Int J Biol Macromol 120:1855–1864

Zaccardi F, Webb DR, Yates T, Davies MJ (2016) Pathophysiology of type 1 and type 2 diabetes mellitus: a 90-year perspective. Postgrad Med J 92(1084):63–69

Zeng X, Ling H, Yang J, Chen J, Guo S (2018) Proteome analysis provides insight into the regulation of bioactive metabolites in *Hericium erinaceus*. Gene 666:108–115

Zhang GQ, Wang YF, Zhang XQ, Ng TB, Wang HX (2010) Purification and characterization of a novel laccase from the edible mushroom *Clitocybe maxima*. Process Biochem 45(5):627–633

Zhang R, Zhao L, Wang H, Ng TB (2014a) A novel ribonuclease with antiproliferative activity toward leukemia and lymphoma cells and HIV-1 reverse transcriptase inhibitory activity from the mushroom, *Hohenbuehelia serotina*. Int J Mol Med 33(1):209–214

Zhang W, Tao J, Yang X, Yang Z, Zhang L, Liu H, Wu K, Wu J (2014b) Antiviral effects of two *Ganoderma lucidum* triterpenoids against enterovirus 71 infection. Biochem Biophys Communn 449(3):307–312

Zhang R, Tian G, Zhao Y, Zhao L, Wang H, Gong Z, Ng TB (2015) A novel ribonuclease with HIV-1 reverse transcriptase inhibitory activity purified from the fungus *Ramaria formosa*. J Basic Microbiol 55(2):269–275

Zhang Y, Hu T, Zhou H, Zhang Y, Jin G, Yang Y (2016) Antidiabetic effect of polysaccharides from *Pleurotus ostreatus* in streptozotocin-induced diabetic rats. Int J Biol Macromol 83:126–132

Zhang L, Hu Y, Duan X, Tang T, Shen Y, Hu B, Liu A, Chen H, Li C, Liu Y (2018a) Characterization and antioxidant activities of polysaccharides from thirteen *Boletus* mushrooms. Int J Biol Macromol 113:1–7

Zhang L, Liu Y, Ke Y, Liu Y, Luo X, Li C, Zhang Z, Liu A, Shen L, Chen H, Hu B (2018b) Antidiabetic activity of polysaccharides from *Suillellus luridus* in streptozotocin-induced diabetic mice. Int J Biol Macromol 119:134–140

Zhang M, Shan Y, Gao H, Wang B, Liu X, Dong Y, Liu X, Yao N, Zhou Y, Li X, Li H (2018c) Expression of a recombinant hybrid antimicrobial peptide magainin II-cecropin B in the mycelium of the medicinal fungus *Cordyceps militaris* and its validation in mice. Microb Cell Fact 17(1):18

Zhao Y, Li X, Chen T, Tang Q, Qiu L, Wang B, Yang Q (2018a) Preparation and antioxidant activity of phosphorylated polysaccharides from *Russula Alutacea* Fr. Ekoloji 27(105):17–22

Zhao X, Zou X, Li Q, Cai X, Li L, Wang J, Wang Y, Fang C, Xu F, Huang Y, Chen B (2018b) Total flavones of fermentation broth by co-culture of *Coprinus comatus* and *Morchella esculenta* induces an anti-inflammatory effect on LPS-stimulated RAW264. 7 macrophages cells via the MAPK signaling pathway. Microbial pathogenesis 125:431–437

Zheng W, Miao K, Liu Y, Zhao Y, Zhang M, Pan S, Dai Y (2010) Chemical diversity of biologically active metabolites in the sclerotia of *Inonotus obliquus* and submerged culture strategies for upregulating their production. Appl Microbiol Biotechnol 87(4):1237–1254

Zheng W, Zhao Y, Zheng X, Liu Y, Pan S, Dai Y, Liu F (2011) Production of antioxidant and antitumor metabolites by submerged cultures of *Inonotus obliquus* cocultured with *Phellinus punctatus*. Appl Microbiol Biotechnol 89(1):157–167

Zhu YC, Wang G, Yang XL, Luo DQ, Zhu QC, Peng T, Liu JK (2010) Agrocybone, a novel bis-sesquiterpene with a spirodienone structure from basidiomycete *Agrocybe salicacola*. Tetrahedr Lett 51(26):3443–3445

Zurga S, Nanut MP, Kos J, Sabotič J (2017) Fungal lectin MpL enables entry of protein drugs into cancer cells and their subcellular targeting. Oncotarget 8(16):26896–26910

Websites Followed

https://www.who.int/news-room/fact-sheets/detail/cancer. Accessed on 16 May 2020

https://www.who.int/news-room/fact-sheets/detail/diabetes. Accessed on 16 May 2020

https://www.who.int/news-room/feature-stories/detail/world-malaria-report-2019. Accessed on 16 May 2020

Chapter 8
Commercialization and Conservation

Macrofungi have great economic value and are traded at world market fetching billions of dollars per annum. This high cost is probably because most species cannot be cultivated but are in great demand for gourmet cooking in many European countries and North America (Yun and Hall 2004; Karwa et al. 2011). As per the retail market, *C. cibarius* and *B. edulis* have estimated 1.67 billion USD and more than 250 million USD per annum, respectively. *A. caesarea* is also in demand in international market. Major countries demanding these species include China, the United States, France, Italy, Spain Canada, and Germany (Hall et al. 2003; Arora and Dunham 2008). Both wild and cultivated macrofungi with use value to mankind contribute tremendously to world economy. The sporocarps are sold fresh or dried in local markets or traded distantly on large scale at national and international level (Gold et al. 2008; Turtiainen and Nuutinen 2012; Bulam et al. 2018). Occurrence of macrofungi in wild is a natural and chance phenomenon. To take advantages of useful macrofungi, these must be made available to the rapidly growing human population and that too at cheaper prices. Moreover, the extraction and purification of fungal metabolites for use as nutraceuticals and pharmaceuticals require large-scale production of fungal biomass. To obtain this meet, several species of useful wild fungi are under cultivation in different regions of the world. The most common mushrooms in cultivation worldwide are *A. bisporus*, *Cordyceps sinensis*, *L. edodes*, *Pleurotus* spp., *Auricularia* spp., and *F. velutipes*, *G. lucidum*, and *Poria cocos*. China leads mushroom production followed by Italy, the United States, Netherlands, and Poland (Feeney et al. 2014; Grimm and Wösten 2018). According to an estimate, mushroom production in China keeps on increasing with 4,826,000–7,786,368 tonnes of mushrooms produced during 2010–2016. Current data shows that the global mushroom market had a value of 35 billion USD in 2015. During the period from 2016 to 2021, this market is expected to grow more by 9.2%, reaching nearly 60 billion USD in 2021 (Raman et al. 2018). The consumption increased from 1 to 4.7 kg of cultivated edible mushrooms per capita during 1997–2013 (Royse et al. 2017). This consumption keeps on increasing in the coming years, resulting in

annual sales going from 34 to 60 billion USD (https://www.zionmarketresearch.
com). Post harvest technology is applied to both the produce of field collection and
cultivation yield of macrofungi for its protection, processing, distribution, marketing, and utilization as per the needs and demands of the consumers (Thakur 2018).
Extracts of many edible/medicinal macrofungi are marketed as DSs for their immunostimulatory and other health-promoting properties (Table 8.1).

Many products own patents too (Chang 2006; Chang and wasser 2018). For
example, nutrient formulations obtained from *A. bisporus*, *L. edodes*, *P. ostreatus*,
and *G. frondosa* (patent application number: WO2007US63984 20070314) are
helpful in the treatment of neurodegenerative diseases and damage from radiations
(Beelman et al. 2007). Food supplement made from *G. frondosa*, *P. eryngii*, and
H. erinaceus (patent application number: CN 101292726 A20081029) reduce blood
sugar and lipid levels (Liu et al. 2008). Ganoderic acid T-amide derivative TLTO-A
(patent application number: CN102219822) causes growth inhibition and apoptosis
of tumor cells (Zhong et al. 2011). A terpenoid (spiroketal) compound from *A. subrufescens* and related *Agaricus* spp. (patent application number: EP 2 468 253 A1)
possesses therapeutic potential against the disorder of having Liver X Receptor
(LXR) agonists activity (Grothe et al. 2012). Mushroom extracts from *Agaricus*
spp., *H. erinaceus*, and *Hypsizygus marmoreus* (patent application number: JP
2012077004A) boost insulin secretion and are taken as health foods for prevention
and curing diabetes mellitus (Takeshi et al. 2012). Several types of mushroom products have been developed in the form of powder, capsules, or tablets using the wild/
cultivated mushrooms or mycelial biomass extracts or spores and extracts of spores
(Pöder 2005; Chang and Wasser 2012) (Fig. 8.1). Medicinal mushrooms have been
used in nearly 400 clinical trials against different ailments. Over 50000 scientific
papers have been published including 15000 patents related to different aspects of
medicinal mushrooms (Wasser 2014). *Ganoderma*, known as "mushroom of immortality," alone holds more than 10000 publications with 7000 related patents. More
than 1000 *Ganoderma* health food products have been certified by Chinese government (Chen et al. 2016). *G. lucidum* is the most commonly used species in the genus
Ganoderma. This is consumed as a functional food such as soup, tea, wine, and
yoghourt in daily life (Dong and Han 2015). Several drugs and food supplements of
G. lucidum are available in China such as *G. lucidum* compound capsules (Chongqing
Taiji Industry (Group) Co. Ltd), syrups (Guizhou Shunjian Pharmaceutical Co.
Ltd), spore powder capsules enriched with Se (Guizhou Lingkangshi Biological
Technology Co. Ltd), *G. lucidum* spore powder (Yunnan Xianghui Pharmaceutical
Co. Ltd), spore powder oil capsules (FGTZ Biotechnology Company), broken
G. lucidum spore powder (Chengdu Dujiangyan Chunsheng Chinese Herbal Pieces
Co. Ltd), and basidiocarp slices (Sichuan Zibo Pharmaceutical Co. Ltd). Further,
many DXN, MAVEX, and other cosmetic products of *Ganoderma* species are avail-
able in market (Hapuarachchi et al. 2018). Another important medicinal mushroom,
I. obliquus (chaga mushroom), products are famous in market such as extracts, powder, chaga tea, capsules, cosmetic products, etc. (Illana-Esteban 2011). In addition
to *G. lucidum* and *I. obliquus*, many more species such as *A. subrufescens*, *C. sinensis*, *L. edodes*, *S. commune*, and *Tremella fuciformis* have been utilized as skin and

Table 8.1 The most popular marketed products of macrofungal extracts with claimed therapeutic potential

Market product	Function	Extract	Reference
BIOIMMUNOGEN PSK-16™ -polysaccharide protein	Immunomodulatory, supports natural killer (NK) cell, T-cell, activity of macrophages and production of cytokine	Mixture of 16 mushroom complex	http://www.qualityahcc.com
Breast-Mate®	Antitumor potential	*Ph. linteus*	Sliva (2010); http://www.swansonvitamins.com
MycoPhyto® Complex	Antiproliferative and cell cycle arrest at the G2/M phase of highly invasive human breast cancer cells, MDA-MB-231	Mixture of mycelia from *A. blazei, C. sinensis, T. versicolor, G. lucidum, G. frondosa,* and *P. umbellatus* and β-1,3-glucan from *Saccharomyces cerevisiae*	Jiang and Sliva (2010)
Mycoformulas Memory	Enhance healthy brain function and nervous system	*H. erinaceus*	http://mycoformulas.com
New Chapter® Life Shield® Immunity	Promotes overall immune health and vitality	*G. lucidum* mycelium and basidiocarps in combination with many other mushrooms	http://www.newchapter.com
ORIVEDA® Coriolus PSP extract	Immunostimulator and adjuvant during anticancer therapy	PSP extract *T. versicolor*	http://www.oriveda.com
Pleuran Imunoglukan P4H® capsules	Immunostimulator	β-Glucan from *P. ostreatus*	http://www.pleuran.com; Bobovcak et al. (2010); Bergendiova et al. (2011)
ReishiMax capsules	Prevents adipocyte differentiation, increases glucose uptake, and activates AMPK	Polysaccharides recovered from *G. lucidum*	Thyagarajan-Sahu et al. (2011)
Coriolus-MRL	Immunostimulator and protects from chronic ailments	Mycelium and primordia of young basidiocarps of *T. versicolor*	http://www.mrlusa.com; Córdoba and Ríos (2012)

(continued)

Table 8.1 (continued)

Market product	Function	Extract	Reference
Breast Defend™	Antiproliferative and metastatic behavior of MDA-MB-231 invasive human breast cancer cells	Extracts of *Trametes versicolor*, *G. lucidum*, and *Ph. linteus*	Jiang et al. (2012)
CV Skinlabs Body Repair Lotion, US; Dr. Andrew Weil for Origins Mega-Mushroom, Skin Relief Face Mask, U.S.; Kat Burki Form Control Marine Collagen, Gel, UK; Tela Beauty Organics Encore Styling, Cream, UK	Wound-healing and anti-inflammatory; boost collagen, improve elasticity, and provide hydration; provide hair with sun protection and prevent color fading	*G. lucidum*	Wu et al. (2016)
Dr. Andrew Weil for Origins Mega-Mushroom, Moisturizing body Cream, US; Estée Lauder, Re-Nutriv Sun Supreme Rescue Serum sun care product, US; Menard Embellir, Night cream, Japan	Skin antiaging; triple-action repair technology to enhance the skin's own natural defenses against the visible effects of sun exposure and sun-stressed skin; eliminate toxins and help repair skin damage associated with overexposure to UV radiation and free radicals	*G. lucidum*	Taofiq et al. (2016)
Capsules	Health supplements	*I. obliquus*, *C. sinensis*, *H. erinaceus*	https://in.pipingrock.com

hair care products. This property is due to the antiaging (firming, lifting, and anti-wrinkle), antioxidant, and moisturizing effects of various mycochemicals such as polysaccharides, phenols, steroids, alkaloids, saponins, etc. present in these mushroom species (Wu et al. 2016). Additionally, some toxigenic species of coprophilous fungi consist of hallucinogenic psychotropic compounds which are used in producing tranquilizers (Griffiths et al. 2016). The advances in biotechnology including DNA barcoding, proteomics, nanotechnology, etc. revolutionize the utilization of macrofungi and commercialization of macrofungal products (Sridhar and Deshmukh 2019).

In order to maintain the increasing demands for fungal material used for cultivation, preservation of mycelia is the most important thing. Generally, mineral oil and distilled water are used for short-term preservation. However, lyophilization and cryopreservation in liquid nitrogen are long-term preservation techniques which inhibit growth and may compromise some metabolic processes (Richter and Bruhn 1989; Croan et al. 1999; Colauto and Linde 2012a, b). During preservation, it is crucial to observe the viability and morphology of the mycelium regularly (Clark

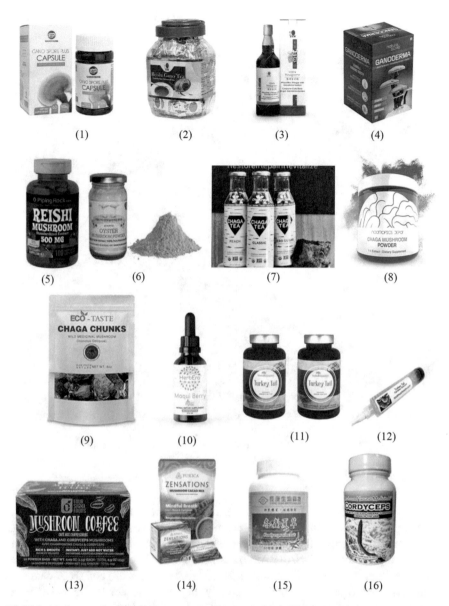

Fig. 8.1 Market products of mushrooms: *Ganoderma lucidum* (Lingzhi/Reishi) capsules (1), tea (2), vinaigrette (3), and capsules (4); *Pleurotus ostreatus* (oyster mushroom) flavoring powder (5); *Inonotus obliquus* (chaga) tea (6), extract (7), alcohol tincture (8), and chunks (9); *Trametes versicolor* (Turkey tail) tincture (10), capsules (11), and powder (12); Arabica coffee with *Inonotus obliquus* and *Cordyceps* mushrooms (13); *H. erinaceus* (Lion's mane) and *Cordyceps* mushrooms with cacao mix (14); *Cordyceps sinensis* capsules (15); *C. militaris* capsules (16); *Phellinus linteus* lotion (17), tea (18), capsules (19), extract (20); *Phellinus igniarius* extract (21) and *Flammulina velutipes* (enokitake) snacks (22). (https://www.amazon.com; https://www.amazon.in; http://asterforce.com; https://www.desertcart.de; https://www.desertcart.in; https://www.dxn2u.com; https://www.exportersindia.com; https://www.21food.com; https://www.foodnavigator-usa.com; https://www.healthysupplies.co.uk; https://herb-era.com; https://mblovesskincare.com; https://www.indiamart.com; http://www.pleuran.com; https://in.pipingrock.com; https://www.qtrove.com; https://www.phytoextractum.com; https://www.tonicology.com; https://www.trianglehealing.com; https://www.tradekorea.com; https://www.shanbally.ie)

(17)   (18)   (19)

(20)   (21)   (22)

Fig. 8.1 (continued)

and Anderson 2004; Marín et al. 2008). It is also important to check the genetic
stability of the mycelium under preservation. Various modern techniques such as
polymerase chain reaction and other methods of DNA analysis such as random
amplified polymorphic DNA (RAPD) markers have been employed to observe the
intraspecific variability (Atienzar et al. 2002; Neves et al. 2005) and DNA damages
and mutations (Lee et al. 2000). Currently, higher fungi are facing great threat
because of habitat loss and overharvesting of wild stocks. Thus, the conservation of
useful fungi, utilized in agro-industrial, medicinal, and commercial purposes is necessary to make them available to the next generations. By creating awareness of
issues posing threat to useful fungi, we wish to promote a sustainable use of these
natural products (Mortimer et al. 2012). In addition to in situ approach, culture collections of fungi representing an ex situ strategy also markedly contribute towards
conservation of macrofungi. Conservation of fungal species diversity and genetic
resources in culture collections provides an essential basis for biotechnological
potential (Bisko et al. 2018). The World Data Center for Microorganisms (WDCM),
founded as the data center for World Federation for Culture Collections (WFCC

established in 1970), collects and keeps the record of worldwide culture collections of microorganisms including fungi. By May 22, 2020, as per the data of WDCM database management system, Culture Collections Information Worldwide (CCINFO), 792 culture collections from 78 countries have been registered. This data also shows that over 3233108 of microorganism cultures have been maintained including 859234 cultures of fungi (http://www.wfcc.info). As a conservation wing of European Mycological Association (EMA) since 2003, the European Council for Conservation of Fungi (ECCF established in 1985) is appreciably improving mycological infrastructure in Europe. The ECCF applies the International Union for Conservation of Nature (IUCN) categories and criteria for the assessment of fungal diversity whenever possible. Therefore, the national Red Lists of fungi in many European countries are prepared at least partly following the IUCN criteria. By 2010, 5500 species of fungi from 33 European countries have red-listed (Dahlberg et al. 2010). A shortlist of fungi comprising 5705 species representing potential candidates for Red List evaluation at the European level is currently available at the ECCF website (http://www.eccf.eu). The red-listing of fungi promotes research initiatives on fungal distribution, population structure, and dynamics and proves helpful to collect data for making conservation decisions and taking actions (Dahlberg and Mueller 2011).

References

Arora D, Dunham SM (2008) A new, commercially valuable Chanterelle species, *Cantharellus californicus* sp. nov., associated with live oak in California, USA. Econ Bot 62(3):376

Atienzar FA, Venier P, Jha AN, Depledge MH (2002) Evaluation of the random amplified polymorphic DNA (RAPD) assay for the detection of DNA damage and mutations. Mutat Res 521(1–2):151–163

Beelman RB, Dubost NJ, Peterson DG, Hausman M (2007) Phytonutrient compositions from mushrooms or filamentous fungi comprising ergothioneine for treatment of neurodegenerative diseases and radiation damage Database: CAPLUS. WO 2007106859 A2 20070920

Bergendiova K, Tibenska E, Majtan J (2011) Pleuran (β-glucan from *Pleurotus ostreatus*) supplementation, cellular immune response and respiratory tract infections in athletes. Eur J Appl Physiol 111(9):2033–2040

Bisko NA, Lomberg ML, Mykchaylova OB, Mytropolska NY (2018) Conservation of biotechnological important species diversity and genetic resource of rare and endangered fungi of Ukraine. Plant & Fungal Research (1):18–27

Bobovcak M, Kuniakova R, Gabriz J, Majtan J (2010) Effect of Pleuran (beta-glucan from *Pleurotus ostreatus*) supplementation on cellular immune response after intensive exercise in elite athletes. Appl Physiol Nutr Metab 35(6):755–762

Bulam S, Üstün NŞ, Pekşen A (2018) Mushroom foreign trade of Turkey in the last decade. In: International congress on engineering and life sciences (ICELIS 2018), pp 779–784

Chang ST (2006) The World mushroom industry: trends and technological development. Int J Med Mushrooms 8(4):297–314

Chang ST, Wasser SP (2012) The role of culinary-medicinal mushrooms on human welfare with a pyramid model for human health. Int J Med Mushrooms 14(2):95–134

Chang ST, Wasser SP (2018) Current and future research trends in agricultural and biomedical applications of medicinal mushrooms and mushroom products (review). Int J Med Mushrooms 20(12):1121–1133

Chen RY, Kang J, Du GH (2016) Construction of the quality control system of *Ganoderma* products. Edib Med Mushrooms 24(6):339–344

Clark TA, Anderson JB (2004) Dikaryons of the basidiomycete fungus *Schizophyllum commune*: evolution in long-term culture. Genetics 167(4):1663–1167

Colauto NB, Linde GA (2012a) Avances sobre el cultivo del "Cogumelo-do-sol" en Brasil. In: Hongos comestibles y medicinales en Iberoamérica: investigación ydesarrollo en un entorno multicultura. Ecosur, Chiapas, pp 121–136

Colauto NB, Linde GA (2012b) Organización y preservación de microorganismos en Brasil. In: Hongos comestibles y medicinales en iberoamérica: investigación y desarrollo en un entorno multicultura. Ecosur, Chiapas, pp 53–68

Córdoba KAM, Ríos AH (2012) Biotechnological applications and potential uses of the mushroom *Tramestes versicolor*. Vitae 19(1):70–76

Croan SC, Burdsall HH Jr, Rentmeester RM (1999) Preservation of tropical wood-inhabiting basidiomycetes. Mycologia 91(5):908–916

Dahlberg A, Mueller GM (2011) Applying IUCN red-listing criteria for assessing and reporting on the conservation status of fungal species. Fungal Ecol 4(2):147–162

Dahlberg A, Genney DR, Heilmann-Clausen J (2010) Developing a comprehensive strategy for fungal conservation in Europe: current status and future needs. Fungal Ecol 3(2):50–64

Dong C, Han Q (2015) *Ganoderma lucidum* (Lingzhi, *Ganoderma*): fungi, algae, and other materials. In: Dietary Chinese herbs chemistry: pharmacology and clinical evidence. Springer, London, pp 759–765

Feeney MJ, Dwyer J, Hasler-Lewis CM, Milner JA, Noakes M, Rowe S, Wach M, Beelman RB, Caldwell J, Cantorna MT, Castlebury LA (2014) Mushrooms and health summit proceedings. J Nutr 144(7):1128S–1136S

Gold MA, Cernusca MM, Godsey LD (2008) A competitive market analysis of the United States shiitake mushroom marketplace. Horttechnology 18(3):489–499

Griffiths RR, Johnson MW, Carducci MA, Umbricht A, Richards WA, Richards BD, Cosimano MP, Klinedinst MA (2016) Psilocybin produces substantial and sustained decreases in depression and anxiety in patients with life-threatening cancer: a randomized double-blind trial. J Psychopharmacol 30(12):1181–1197

Grimm D, Wösten HAB (2018) Mushroom cultivation in the circular economy. Appl Microbiol Biotechnol 102(18):7795–7803

Grothe T, Stadler M, Köpcke B, Roemer E, Bitzer J, Wabnitz P (2012) Terpenoid spiro ketal compounds with LXR agonists activity, their use and formulations with them. EP2468253 A1

Hall IR, Wang Y, Amicucci A (2003) Cultivation of edible ectomycorrhizal mushrooms. Trends Biotechnol 21(10):433–438

Hapuarachchi KK, Elkhateeb WA, Karunarathna SC, Cheng CR, Bandara AR, Kakumyan P, Hyde KD, Daba GM, Wen TC (2018) Current status of global *Ganoderma* cultivation, products, industry and market. Mycosphere 9(5):1025–1052

Illana-Esteban C (2011) Interés medicinal del "Chaga" (*Inonotus obliquus*). Bol Soc Micol 35:175–185

Jiang J, Sliva D (2010) Novel medicinal mushroom blend suppresses growth and invasiveness of human breast cancer cells. Int J Oncol 37(6):1529–1536

Jiang J, Thyagarajan-Sahu A, Loganathan J, Eliaz I, Terry C, Sandusky GE, Sliva D (2012) BreastDefend™ prevents breast-to-lung cancer metastases in an orthotopic animal model of triple-negative human breast cancer. Oncol Rep 28(4):1139–1145

Karwa A, Vamma A, Rai M (2011) Edible ectomycorrhizal fungi: cultivation, conservation and challenges. In: Diversity and biotechnology of ectomycorrhizae. Springer, Berlin/Heidelberg, pp 429–453

Lee YK, Chang HH, Kim JS, Kim JK, Lee KS (2000) Lignocellulolytic mutants of *Pleurotus ostreatus* induced by gamma-ray radiation and their genetic similarities. Radiat Phys Chem 57(2):145–150

Liu Z, Cang L, Chen C, Zhou G (2008) Method for preparing gold needle mushroom canned food for decreasing blood sugar and adjusting blood fat Database: CAPLUS. CN 101292726 A 20081029

Marín S, Cuevas D, Ramos AJ, Sanchis V (2008) Fitting of colony diameter and ergosterol as indicators of food borne mould growth to known growth models in solid medium. Int J Food Microbiol 121(2):139–149

Mortimer PE, Karunarathna SC, Li Q, Gui H, Yang X, Yang X, He J, Ye L, Guo J, Li H, Sysouphanthong P (2012) Prized edible Asian mushrooms: ecology, conservation and sustainability. Fungal Divers 56(1):31–47

Neves MA, Kasuya MCM, Araújo EF, Leite CL, Camelini CM, Ribas LCC, Mendonça MM (2005) Physiological and genetic variability of commercial isolates of culinary-medicinal mushroom *Agaricus brasiliensis* S. Wasser et al. (Agaricomycetideae) cultivated in Brazil. Int J Med Mushrooms 7(4):575–585

Pöder R (2005) The Ice man's fungi: facts and mysteries. Int J Med Mushrooms 7(3):357–359

Raman J, Lee S-K, Im J-H, Oh M-J, Oh Y-L, Jang K-Y (2018) Current prospects of mushroom production and industrial growth in India. J Mushrooms 16(4):239–249

Richter DL, Bruhn JN (1989) Revival of saprotrophic and mycorrhizal basidiomycete cultures from cold storage in sterile water. Can J Microbiol 35(11):1055–1060

Royse DJ, Baars J, Tan Q (2017) Current overview of mushroom production in the world. In: Edible and medicinal mushrooms: technology and applications. Wiley, Hoboken, pp 5–13

Sliva D (2010) Medicinal mushroom *Phellinus linteus* as an alternative cancer therapy. Exp Ther Med 1(3):407–411

Sridhar KR, Deshmukh SK (eds) (2019) Advances in macrofungi: diversity, ecology and biotechnology. CRC Press

Takeshi I, Hiroshi H, Satoshi I, Aya K (2012) Mushroom extracts from Agaricus , Hericium erinaceum, and Hypsizigus marmoreus as insulin secretion stimulators and health foods for prevention and therapy of diabetes mellitus. JP 2012077004A

Taofiq O, González-Paramás AM, Martins A, Barreiro MF, Ferreira IC (2016) Mushrooms extracts and compounds in cosmetics, cosmeceuticals and nutricosmetics – a review. Ind Crop Prod 90:38–48

Thakur MP (2018) Advances in post-harvest technology and value additions of edible mushrooms. Indian Phytopathol 71(3):303–315

Thyagarajan-Sahu A, Lane B, Sliva D (2011) ReishiMax, mushroom based dietary supplement, inhibits adipocyte differentiation, stimulates glucose uptake and activates AMPK. BMC Complement Altern Med 11(1):1–4

Turtiainen M, Nuutinen T (2012) Evaluation of information on wild berry and mushroom markets in European countries. Small-scale For 11(1):131–145

Wasser SP (2014) Medicinal mushroom science: current perspectives, advances, evidences, and challenges. Biomed J 37(6):345–356

Wu Y, Choi MH, Li J, Yang H, Shin HJ (2016) Mushroom cosmetics: the present and future. Cosmetics 3(3):22

Yun W, Hall IR (2004) Edible ectomicorrhizal mushrooms: challenges and achievements. Can J Bot 82(8):1063–1073

Zhong J, Liu R, Li Y (2011) Ganoderic acid T amide derivative TLTO-A and synthetic method and application thereof Database: CAPLUS. CN 102219822 A 20111019

Websites Followed

http://asterforce.com
http://mycoformulas.com
http://www.eccf.eu. Accessed on 22 May 2020
http://www.mrlusa.com
http://www.newchapter.com
http://www.oriveda.com
http://www.pleuran.com
http://www.qualityahcc.com
http://www.swansonvitamins.com
http://www.wfcc.info. Accessed on 22 May 2020
https://herb-era.com
https://in.pipingrock.com
https://mblovesskincare.com
https://www.21food.com
https://www.amazon.com
https://www.amazon.in
https://www.desertcart.de
https://www.desertcart.in
https://www.dxn2u.com
https://www.exportersindia.com
https://www.foodnavigator-usa.com
https://www.healthysupplies.co.uk
https://www.indiamart.com
https://www.phytoextractum.com
https://www.qtrove.com
https://www.shanbally.ie
https://www.tonicology.com
https://www.tradekorea.com
https://www.trianglehealing.com
https://www.zionmarketresearch.com

Chapter 9
Outlook and Future Prospects

The major step before consumption of the collected macrofungal species is correct identification. Misidentification of macrofungi often leads to the consumption of toxigenic species causing adverse health effects and sometimes may prove lethal. Without correct scientific nomenclature of the collection material, further studies have no validity. For example, in earlier pharmacological investigations, most of the species reported as Lingzhi or Reishi (*G. lucidum*) were mistakenly identified. *G. lucidum* represents a taxon-linneon or species-complex, and the future subdivision of this needs caution (Wasser et al. 2006). Therefore, due to misidentification, the publications, patents, and products are also at risk. Classical methods include the study of morphology (color, size, shape, etc.) and microscopic examination (hyphae, spores, sporing structures, sterile elements, etc.) of the specimen. Field guides, monographs, and identification keys bearing photographs and detailed descriptions of species are also beneficial in identification. Preserving wild specimen in different herbaria is greatly encouraged to physically compare and identify the specimens collected later. In addition to classical taxonomic methods, advances in electron microscopy and molecular techniques bring significant improvements and rapidity in the identification, nomenclature, and classification of the specimens (Abdel-Azeem et al. 2018; Wu et al. 2018). The gathered information from different sources is compiled into various databases and is made available online on various websites such as Mycobank, Species Fungorum, MyCoPortal, etc. A vast diversity of macrofungi exists on earth with continuous additions of species from different parts of the world. Still there are some regions left unexplored or are underexplored. More field work is needed to explore and document macrofungal diversity, particularly in the tropics utilizing morphological and molecular identification techniques (Lopez-Quintero et al. 2012; Ambrosio et al. 2015; Iqbal et al. 2016). Macrofungi have long history of use in human life as is evident from ethnomycology. In addition to cultural significance, the information pertaining to the edibility, medicinal value, or toxicity of macrofungi can be gathered by conducting ethnomycological surveys to different parts of the world (Yamin-Pasternak, 2011; Ingvar 2018). The ethnomycological data about the uses of macrofungi in human health need scientific valorization which is possible through the analysis of mycochemical composition and evaluation of biological activities. This has become possible with use of bio-technology involving genomics, proteomics, and other modern-day techniques such as WDXRF, chromatography, and mass spectrometry, e.g., GC/MS, LC/MS, etc. (Lozano et al. 2012; Lau et al. 2014; Parmar and Kumar 2015; Ham et al. 2020; Shao et al. 2020). Besides aiding in identification, biotechnology makes it possible to generate complete mycochemical profiles, to elucidate the structures of bioactive compounds, and to evaluate the pharmacological activities of macrofungal species (Liu et al. 2017; Dong et al. 2018). However, numerous species of macrofungi still need scientific evaluation with respect to their mycochemical composition and pharmacological potential. Further advances in medical mycology are the need of the hour to make the clinical use of macrofungi in the treatment of various ailments. Empirical evidences may prove helpful to decide the edibility of a species. The scientific literature also proves as the best source of advice. The ingestion of fresh mushrooms rich in health boosting nutrients and other bioactive metabolites is beneficial, e.g., it enhances the serum levels of anti-β-glucan antibodies. As per the opinion of the Ohno Group from Japan, mushrooms when taken fresh would provide defense from pathogenic fungi in a better way (Ishibashi et al. 2005). Currently, much attention is paid to develop mushroom biomass or extracts as DSs or functional foods or prebiotics (indigestible β-glucans). However, a number of significant queries arise with establishing DSs and medicinal mushroom products concerning their safety, standardization, regulation, efficacy, and mechanism of action. However, this is a pity that even today standardization of DSs and medicinal mushroom products is in its early stages including the insufficient knowledge of the bioactive constituents and consequences of consumption of DSs. The standards and protocols employed to produce and test the medicinal mushroom products are not well recognized. The lack of proper standards and protocols results into adulterated and less durable mushroom products making them markedly different in composition and affectivity. It is unclear whether the health effect is produced by a single component or is the outcome of synergistic action of several ingredients. Moreover, insufficient data is available to decide which components possess better potential – those recovered from sporocarps or from submerged mycelia powder versus extracts. Other questions which frequently arise include whether the simple dried sporocarps or mycelia powders are more effective compared to aqueous, alcoholic, or hydroalcoholic extracts. Moreover, between crude extracts and isolated fractions, which one is more effective and safe? What is the role of low molecular weight compounds present in medicinal macrofungi? What is more effective, the combination of bioactive component-enriched biomass or extracts of 2–10 different species in a single pill or one species in one pill? Is it possible to access the potential of different mushroom products when mixing in several species in single product ("shotgun" approach)? Since the mushroom products can be cytokine stimulants, at what age these can be safe to administer in young children as their immune system is immature? The other controversial thing is what dosages are safe and affective especially during pregnancy and nursing? In several clinical trials, six capsules (three capsules twice a day or two capsules thrice a day) of 500–1000 mg each (biomass or extracts) have been accepted as the affective dosage of medicinal mushroom formulations. As per the traditional Chinese medicine, the standard dose of tablets, capsules, liquid extracts, etc. made from dried sporocarps or biomass must be 100–150 g a day of fresh mushroom material (Wasser 2014). The high molecular weight compounds, especially polysaccharides of medicinal macrofungi are used for drug development. However, these cannot be synthesized, and their production is confined to isolation from sporocarps/cultured mycelium or cultured broth which is an expensive approach (Papaspyridi et al. 2018). Therefore, an attempt should be made for drug development from low molecular weight compounds (Zaidman et al. 2005). Furthermore, the adulteration of mushroom products causes adverse health impacts such as nephropathy, acute hepatitis, coma, and fever. For example, the production of pure β-glucans is difficult, and 90–95% of β-glucan in the market are considered adulterated (Deng et al. 2009; Wasser 2011). Therefore, we must pay attention to adopt regulations, standards, and practices most valuable in the quest for human health promotion. Research should lay emphasis on improvements in cultivation technology, and attempts should be made for successful cultivation of useful macrofungi which are not under cultivation yet. The protection of intellectual properties (IPs) of genetic resources of medicinal macrofungi needs more attention. These are used and exploited by the pharmaceutical, cosmetic, agricultural, food, enzyme, and chemical industries. We should educate the public and make them aware of the use value of macrofungi in boosting human health. The most important above all is the conservation of the inexpensive natural treasure of macrofungi for sustainable use.

References

Abdel-Azeem AM, Blanchette RA, Held BW (2018) New record of *Chaetomium grande* Asgari & Zare (Chaetomiaceae) for the Egyptian and African mycobiota. Phytotaxa 343(3):283–288

Ambrosio E, Lancellotti E, Brotzu R, Salch H, Franceschini A, Zotti M (2015) Assessment of macrofungal diversity in Silver Fir plantation in Sardinia (Italy) using a standardized sampling procedure. Micol Ital 44(1):1–17

Deng G, Lin H, Seidman A, Fornier M, D'Andrea G, Wesa K, Yeung S, Cunningham-Rundles S, Vickers AJ, Cassileth BA (2009) A phase I/II trial of a polysaccharide extract from Grifola frondosa (Maitake mushroom) in breast cancer patients: immunological effects. J Cancer Res Clin Oncol 135(9):1215–1221

Dong M, Qin L, Xue J, Du M, Lin S-Y, Xu X-B, Zhu BW (2018) Simultaneous quantification of free amino acids and 5′-nucleotides in shiitake mushrooms by stable isotope labeling-LC-MS/MS analysis. Food Chem 268(1):57–65

Ham TH, Lee Y, Kwon SW, Jang MJ, Park YJ, Lee J (2020) Increasing coverage of proteome identification of the fruiting body of *Agaricus bisporus* by Shotgun proteomics. Foods 9(5):632

Ingvar I (2018) Ethnomycological Notes on *Haploporus odorus* and other polypores in Northern Fennoscandia. J North Stud 12(1):73–91

Iqbal AM, Vidyasagaran K, Ganesh PN (2016) New records of polypores (Basidiomycota: Aphyllophorales) from the southern Western Ghats with an identification key for polypores in peechi-vazhani Wildlife sanctuary, Kerala, India. J Threat Taxa 8(9):9198–9207

Ishibashi KI, Dogasaki C, Iriki T, Motoi M, Kurone YI, Miura NN, Adachi Y, Ohno N (2005) Anti-β-glucan antibody in Bovine sera. Int J Med Mushrooms 7(4):513

Lau CC, Abdullah N, Shuib AS, Aminudin N (2014) Novel angiotensin I-converting enzyme inhibitory peptides derived from edible mushroom *Agaricus bisporus* (J.E. Lange) Imbach identified by LC–MS/MS. Food Chem 148(1):396–401

Liu K, Xiao X, Wang J, Chen CYO, Hue H (2017) Polyphenolic composition and antioxidant, antiproliferative and antimicrobial activities of mushroom *Inonotus sanghuang*. LWT-Food Sci Technol 82(1):154–161

Lopez-Quintero CA, Straastsma G, Franco-Molano AE, Boekhout T (2012) Macrofungal diversity in Colombian Amazon forests varies with regions and regimes of disturbance. Biodivers Conserv 21:2221–2243

Lozano A, Martínez-Uroz MA, Gómez-Ramos MJ, Gómez-Ramos MM, Mezcua M, Fernández-Alba AR (2012) Determination of nicotine in mushrooms by various GC/MS and LC/MS-based methods. Anal Bioanal Chem 402:935–943

Papaspyridi L-M, Zerva A, Topakas E (2018) Biocatalytic synthesis of fungal β-glucans. Catalysts 8(7):274

Parmar R, Kumar D (2015) Study of chemical composition in wild edible mushroom *Pleurotus cornucopiae* (Paulet) from Himachal Pradesh, India by using fourier transforms infrared spectrometry (FTIR), gas chromatography-mass spectrometry (GCMS) and X-ray fluorescence (XRF). Biol Forum Int J 7(2):1057–1066

Shao Y, Guo H, Zhang J, Liu H, Wang K, Zuo S, Xu P, Xia Z, Zhou Q, Zhang H, Wang X, Chen A, Wang Y (2020) The genome of the medicinal macrofungus sanghuang provides insights into the synthesis of diverse secondary metabolites. Front Microbiol 10:3035

Wasser SP (2011) Current findings, future trends, and unsolved problems in studies of medicinal mushrooms. Appl Microbiol Biotechnol 89(5):1323–1332

Wasser SP (2014) Medicinal Mushroom Science: current perspectives, advances, evidences, and challenges. Biomed J 37(6):345–356

Wasser SP, Zmitrovich IV, Didukh MY, Spirin WA, Malysheva VF (2006) Morphological traits of *Ganoderma lucidum* complex highlighting *G. tsugae* var. *jannieae*: the current generalization. ARA Gantner Verlag K-G, Ruggell

Wu G, Lee SML, Horak E, Yang ZL (2018) *Spongispora temasekensis*, a new boletoid genus and species from Singapore. Mycologia 110(5):919–929

Yamin-Pasternak S (2011) Ethnomycology: fungi and mushrooms in cultural entanglements. In: Ethnobiology. Wiley, Hoboken, pp 213–230

Zaidman BZ, Yassin M, Mahajna J, Wasser SP (2005) Medicinal mushroom modulators of molecular targets as cancer therapeutics. Appl Microbiol Biotechnol 67:453–468

Chapter 10
Concluding Remarks

Our literature survey reveals that majority of the health benefitting taxa of macro-fungi chiefly belong to *Agaricomycotina* followed by *Pezizomycotina* with domi-nancy of species in *Agaricomycetes*. In the present work, we made an effort to shed light on the health-promoting efficacies of the edible and medicinal macrofungi and discussed these species following the classification of Kirk et al. (2008) current names of macrofungi as per Mycobank database (www.mycobank.org). Macrofungi are represented by 1,250,000 specimens across the globe (Thiers and Halling 2018). This data keeps on changing with the underexplored and unexplored ecogeographic zones. *Macromycetes* exhibit variations in structure and reproduction, ecology, and edibility. These species are capable of growing on a broad spectrum of substrates such as various types of soil rich in moisture and humus, litter, animal dung, dead decaying wood and organic matter, termite mounds, live trees, and a few even on plastic debris (Brunner et al. 2018; Karun et al. 2018). These species play several pivotal functions in the forest ecosystems such as in cycling of nutrients, as ecosys-tem scavengers, as pathogens, and as symbionts (Bajpai et al. 2019; Copoţ and Tănase 2019). Macrofungi have wide distribution range occurring in various tropi-cal, subtropical, temperate, and alpine forests and even in deserts of different coun-tries across Europe, Africa, America, Asia, and Australia (Tripathi et al. 2017). *Macromycetes* (edible/medicinal) have marked significance as human health-promising agents. Edible species in the genera *Agaricus*, *Auricularia*, *Betula*, *Cordyceps*, *Morchella*, *Pleurotus*, *Russula*, *Terfezia*, etc. are eaten as food, while medicinal species placed in the genera *Fomes*, *Ganoderma*, *Inonotus*, *Phellinus*, *Trametes*, etc. are utilized to treat different kinds of diseases in different parts of the world as is evidenced by ethnomycology. The sporocarps of wild edible species of macrofungi are eaten after cooking following different recipes across the world. For medicinal use, the sporocarps of medicinal species are used as whole, in the form of powder or paste. Additionally, some species are toxigenic and accidental consump-tion of such species cause several precarious health effects such as vomiting, nau-sea, stomachache, gastroenteritis, diarrhea, hepatotoxicity, nephrotoxicity,

© The Editor(s) (if applicable) and The Author(s), under exclusive license
to Springer Nature Switzerland AG 2020
U. Azeem et al., *Fungi for Human Health*,
https://doi.org/10.1007/978-3-030-58756-7_10

neurotoxicity, etc. and in certain cases may prove fatal (Erguven et al. 2007; Vişneci et al. 2019). Therefore, correct identification is prerequisite for the consumption of these species. However, various toxins obtained from the toxigenic species, e.g. *A. xanthodermus*, possess medicinal properties (Özaltun and Sevindik 2020). Macrofungi are enriched with diverse kinds of nutrients such as carbohydrates, sugars, proteins, fats, fatty acids, amino acids, minerals, and vitamins responsible for their food value. The medicinal mushrooms possess high as well a low molecular weight bioactive constituents. High molecular weight compounds mainly include polysaccharides (β-glucans) and proteins, whereas low molecular weight compounds are secondary metabolites, e.g., terpenoids, phenols, tannins, flavonoids, steroids, alkaloids, etc. All these mycochemicals contribute towards pharmacological activities of mushrooms such as antioxidant, antitumor, antidiabetic, antibacterial, antifungal, anti-inflammatory, immunomodulatory, antimalarial, antiviral, and neurodegenerative. Evolution of modern-day techniques such as electron microscopy, WDXRF, GC-MS, LC-MS, etc. revolutionizes the identification, analysis of mycochemical composition, and evaluation of pharmacological potential of macrofungi. The mycelial biomass or extracts of health-promoting edible/medicinal macrofungi are being developed as DSs/functional foods or prebiotics. Several medicinal macrofungal products are available in the market as health supplements in the form of tablets, capsules, powder, or extract, as cosmetics (skin, hair products), tea, anti-fatigue and immunostimulatory agents, etc. There are various regulatory authorities such as WHO, FAO, and FDA to manage the use and ensure the quality of these commercial products. The wild mushrooms grow naturally, and their growth and occurrence in nature is a chance phenomenon. To meet the growing needs of higher fungi as food, in medicine and industry, cultivation is the best option. Several species of macrofungi with use value to man are in cultivation worldwide with China leading in this industry. *A. bisporus*, *C. sinensis*, *L. edodes*, *Pleurotus* spp., *Auricularia* spp., and *F. velutipes*, *G. lucidum*, and *P. cocos* represent the commonly cultivated species in the world. In spite of these inestimable benefits of macrofungi, these are often overlooked and often kept out of conservation plans. Natural calamities like forest fire, floods, tornadoes, etc. disturb the structure and function of ecosystem which is a home to diverse forms of life including macrofungi (Salo and Kouki 2018; Ford et al. 2018). Anthropogenic activities such as deforestation, industrialization, and urbanization are responsible for climatic change which in turn bring changes in the diversity and distribution of macrofungi (Lees and Pimm 2015; Wagensommer et al. 2018). Because of overexploitation, anthropogenic activities, and climate change, wild macrofungi are facing threat of extinction, and several taxa are red listed by IUCN (http://www.eccf.eu). Therefore, for sustainable use of wild macrofungi and to maintain the genetic stability in culture collections, several in situ and ex situ conservation efforts are made regularly by IUCN and other national and international organizations.

References

Bajpai A, Rawat S, Johri BN (2019) Fungal diversity: global perspective and ecosystem dynamics. In: Microbial diversity in ecosystem sustainability and biotechnological applications. Springer, Singapore, pp 83–113

Brunner I, Fischer M, RuÈthi J, Stierli B, Frey B (2018) Ability of fungi isolated from plastic debris floating in the shoreline of a lake to degrade plastics. PLoS ONE 13(8):1–14

Copoţ O, Tănase C (2019) Lignicolous fungi ecology-biotic and abiotic interactions in forest ecosystems. Memoirs of the scientific sections of the Romanian academy, 42

Erguven M, Yilmaz O, Deveci M, Aksu N, Dursun F, Pelit M, Cebeci N (2007) Mushroom poisoning. Indian J Pediatrics 74:51–56

Ford SA, Kleinman JS, Hart JL (2018) Effects of wind disturbance and salvage harvesting on macrofungal communities in a *Pinus* woodland. Forest Ecol Manag 407:31–46

Karun NC, Bhagya BS, Sridhar KR (2018) Biodiversity of macrofungi in Yenepoya campus, Southwest India. Microb Biosyst 3(1):1–11

Kirk PM, Cannon PF, Minter DW, Stalpers JA (2008) Ainsworth and Bisby's dictionary of the fungi, 10th edn. CABI, Wallingford

Lees AC, Pimm SL (2015) Species, extinct before we know them? Curr Biol 25(5):177–180

Özaltun B, Sevindik M (2020) Evaluation of the effects on atherosclerosis and antioxidant and antimicrobial activities of *Agaricus xanthodermus* poisonous mushroom. Eur Res J:1–7

Salo K, Kouki J (2018) Severity of forest wildfire had a major influence on early successional ectomycorrhizal macrofungi assemblages, including edible mushrooms. Forest Ecol Manag 415:70–84

Thiers BM, Halling RE (2018) The macrofungi collection consortium. Appl Plant Sci 6(2):1–7

Tripathi NN, Singh P, Vishwakarma P (2017) Biodiversity of macrofungi with special reference to edible forms: a review. J Indian Bot Soc 96(3):144–187

Vişneci EF, Acar D, Özdamar EN, Güven M, Patat M (2019) Mushroom poisoning cases from an emergency department in Central Anatolia: comparison and evaluation of wild and cultivated mushroom poisoning. Eurasian J Emerg Med 18(1):28–33

Wagensommer RP, Bistocchi G, Arcangeli A, Rubini A, Perini C, Venanzoni R, Angelini P (2018) An assessment of red list data for the Pezizomycotina (Ascomycota): Umbria (Italy) as a test case. Plant Biosyst 152(6):1329–1337

Websites Followed

http://www.eccf.eu
www.mycobank.org

Printed in the United States
by Baker & Taylor Publisher Services